Population Ecology

Population Ecology

First Principles

John H. Vandermeer
and
Deborah E. Goldberg

PRINCETON UNIVERSITY PRESS   PRINCETON AND OXFORD

Copyright © 2003 by Princeton University Press
Published by Princeton University Press, 41 William Street, Princeton, New Jersey 08540

In the United Kingdom: Princeton University Press, 3 Market Place, Woodstock, Oxfordshire OX20 1SY

Library of Congress Cataloging-in-Publication Data

Vandermeer, John H.
    Population ecology : first principles / John H. Vandermeer and Deborah E. Goldberg.
    p. cm.
    Includes bibliographical references and index.
    ISBN 0-691-11440-4—ISBN 0-691-11441-2 (pbk.)
    1. Population biology—Mathematical models.   2. Ecology—Mathematical models.
    I. Goldberg, Deborah Esther.   II. Title.

    QH352 .V36 2003
    577.8'8—dc21   2002035478

British Library Cataloging-in-Publication Data is available

This book has been composed in Palatino with Albertus MT Light and Regular Display

Printed on acid-free paper. ∞

ww.pupress.princeton.edu

Printed in the United States of America

1   3   5   7   9   10   8   6   4   2

To Benjamin, Jason, and Jaime,
who are so much more than just our contribution to global r

Contents

CHAPTER 4

A Closer Look at the "Dynamics" in Population Dynamics

CHAPTER 5

Patterns in Space and Metapopulations

CHAPTER 6

Predator–Prey (Consumer–Resource) Interactions

List of Figures

List of Tables

P r e f a c e

The average ecology student today faces some daunting literature. The legacy of the late sixties and early seventies is a fascinating, yet sometimes perplexing, collection of theoretical approaches that have transformed the field dramatically. A course in ecology today is as likely to contain complicated differential equations as descriptions of life histories, a dramatic change from the situation before the 1960s, when mathematics was little mentioned in ecology. And observational studies as much as experiments now rely on predictions from abstract theory. Ecology has become a remarkably exciting discipline principally because of this burgeoning theoretical superstructure.

Along with the momentum derived from this paradigm shift, we also see some frustration as students confront this literature. In our experience, this frustration is frequently derived from a less than adequate appreciation for what have become the fundamental quantitative principles of ecology. This book is an attempt to present these fundamentals in as heuristic a way as possible.

Precisely what are those fundamentals? Attempts at summarizing the underlying quantitative framework for the entire field of ecology can be frustrating. Indeed, the content of a course in "general ecology" at a U.S. university may vary considerably. The choice and arrangement of subject matter are only loosely correlated among recent texts. This variability is, we believe, the consequence of the youthfulness of our science. This lack of consistency may seem frustrating to some, but it is a source of excitement for most ecologists.

Nevertheless, there is one subdiscipline of ecology that seems to have developed a canon. Virtually all courses that contain "population ecology" in their title cover essentially the same material. In this one corner of the field, ecologists seem to agree on what constitutes the basic subject matter. Any survey of contemporary literature in ecology will uncover one or more of the basic ideas of population ecology underlying almost every published study.

This text attempts to present these first principles of population ecology. It is intended as a text for advanced undergraduate and beginning graduate students and focuses on the analytical details of the basic subject matter. It is not at all intended to be an introduction to the literature. Individual applications are chosen as examples of the models and techniques presented and are not intended to be the best or the best-known studies in the literature: they are simply examples. Indeed, a review of the literature of population ecology would fill several volumes and would likely not serve much purpose anyway. This book is not such an effort.

Although the subject matter of population ecology, and thus this book, is highly mathematical, the material is presented in such a way that only introductory calculus and a basic understanding of linear algebra are prerequisites. For students who took calculus long ago, a basic understanding of the nature of a derivative and an integral plus the ability to differentiate simple functions (polynomials and exponentials) are really all that is required to fully appreciate this basic material. On the other hand, many students of ecology do not come to the field with a proper background in the basic operations of linear algebra. For these students we include an appendix at the end of chapter 3.

The organization of the book is based on chapter 1 as a foundation. With the exception of chapter 8, which depends to some extent on material from chapter 6, the rest of the chapters could stand alone, after chapter 1 has been read. Chapter 2 is an example of the application of chapter 1 to some theoretical issues involving life histories. A much more complete presentation of this material can be found in Stearns 1992. Chapters 3 and 4 represent some complications that enter into the application of the principles of chapter 1. Chapter 3 covers the simplest aspects of structured models, emphasizing age and stage models. Elaborations of this introductory material can be found in Caswell's excellent text (Caswell 2001). Chapter 4 introduces nonlinearities to the basic models, a subject that has witnessed an explosion in the literature, although most of the treatments are difficult mathematically. Two references that should be accessible to most ecology students are the two-volume set by Jackson (1991) and the less inclusive but perhaps simpler text by Aligood, Sauer, and Yorke (1996). Chapter 5 deals both with statistical descriptions of spatial aggregation of individuals and populations and with population dynamic models that incorporate spatial information, especially metapopulation dynamics. Spatial statistics is a burgeoning field, especially with the recent growth in the use of geographic information systems (GIS). Our treatment is elementary, and students interested in this subject need to consult more sophisticated treatments (e.g., Diggle 1983). Metapopulation theory is introduced in its traditional form strictly as a mean field theory. More detailed elaborations of the concept can be found in Hanski's recent monograph (Hanski 1999).

Finally, chapters 6, 7, and 8 are introductions to two-species interactions. Chapter 6 deals with positive–negative interactions generally, specifically formulated as predator–prey interactions. More sophisticated treatments of this material can be found in the many applications of predator–prey theory in the literature, but we know of no particular text that treats advanced predator–prey theory as such. Useful applications can be found in Hawkins and Cornell 1999. Epidemiology is really the conceptual application of predator–prey theory to microparasites, but the framework that is used is basically that of metapopulations, as discussed in chapter 7. The standard reference on this material has become Anderson and May 1991. Chapter 8 deals with interspecific competition. More complex elaborations of the material presented in chapter 8 can be found in Tilman 1982, MacArthur 1970, and Chesson 1990.

All of the material in this book has been presented to four groups of graduate and advanced undergraduate students at the University of Michigan. We owe a debt of gratitude to those students for their contributions to improvements that led to the present text. We also thank our colleagues Mark Wilson, who contributed substantially to initial drafts, and Mercedes Pascual, who was kind enough to use a draft of it in her class on population ecology.

Population Ecology

1
CHAPTER

Elementary Population Dynamics

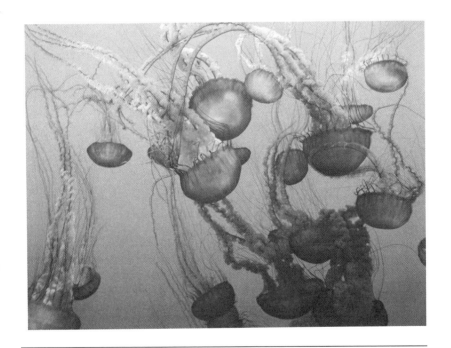

In many contexts it is important to understand the characteristics of single populations of organisms. A wildlife manager, for example, needs to predict what the density of a population of deer or cod would be under different management plans. Or an agronomist may wish to know the yield of a population of maize plants when planted at a particular density. In more theoretical applications, we are interested in knowing, for example, the rate at which a population changes its density in response to selection pressure. These topics are typical of the field called population ecology.

The unit of analysis is, not surprisingly, the population, a concept that is at once simple and complicated. The simple idea is that a population is a collection of individuals. But, as most ecologists intuitively know, the idea of a population is considerably more complex when one deals with the sort of real-life examples mentioned above. To know what size limits one should place on catch for a fish species, one must know not only how many fish are in the population but also the size distribution of that population and how that distribution is related to the population's overall reproduction. To decide when to take action on the emergence of pest species in forests or farms, one must know the distribution of individuals within life stages. In the determination of whether a species is threatened with extinction, its distribution in space and the amount of movement among subpopulations (i.e., metapopulation dynamics) are far more important than simply its numerical abundance. And, to cite the most cited example, the absolute abundance of the human population has little to do with anything of interest com-

---

Jellyfish sometimes form large populations that grow exponentially, at least for a while.

pared with the activities undertaken by the members of that population.

Thus, the subject of population ecology can be very complicated. But, as we do in any science, we begin by assuming that it is simple. We eliminate the complications, make simplifying assumptions, and try to develop general principles that might form a skeleton upon which the flesh of real-world complications might meaningfully be attached. This chapter covers the first two essential ideas of that skeleton: density independence and density dependence.

Density Independence: The Exponential Equation

It is surprising how quickly a self-reproducing phenomenon becomes big. The classic story goes like this: Suppose you have a lake with some lily pads in it and suppose each lily pad replicates itself once a week. If it takes a year for half the lake to become covered with lily pads, how long will it take for the entire lake to become covered? If one does not think too long or too deeply about the question, the quick answer seems to be about another year. But a moment's reflection retrieves the correct answer, only one more week.

This simple example has many parallels in real-world ecosystems. A pest building up in a field may not seem to be a problem until it is too late. A disease may seem much less problematical than it really is. The simple problem of computing the action threshold (the density a pest population must reach before you have to spray pesticide) requires the ability to predict a population's size on the basis of its previous behavior. If half the plants have been attacked within 3 months, how long will it be before they are all attacked?

To understand even the extremely simple example of the lily pads, one constructs a mathematical model, usually quite informally in one's head. If all the lily pads on the pond replicate themselves once a week, then, in a pond half-filled with lily pads, each one of those lily pads will replicate in the next week and thus the pond will be completely filled up. To make the solution to the problem general we say the same thing, but instead of labeling the entities lily pads, we call them something general, say organisms. If organisms replicate once a week and the environment is half full, it will take only one week for it to become completely full. Implicitly, the person who makes such a statement is saying out loud the following equation:

$$N_{t+1} = 2N_t \qquad (1)$$

N is the number of organisms, in this case lily pads. Instead of t(time), say this week, and instead of $t+1$, say next week, and equation 1 is simply "the number of lily pads next week is equal to twice the number this week."

Of course, writing down equation 1 is no different than making any of the statements that were made previously about it. But by making it explicitly a mathematical expression, we bring to our potential use all the machinery of formal mathematics. And that is actually good, even though beginning students sometimes do not think so.

Using equation 1, we can develop a series of numbers that reflect the changes of population numbers over time. For example, consider a population of herbivorous insects: if each individual produces a single offspring once a week, and those offspring mature and also produce an offspring within a week, we can apply equation 1 to see exactly how many individuals will be in the population at any point in time. Beginning with a single individual we have, in subsequent weeks, 2, 4, 8, 16, 32, 64, 128, and so on. If we change the conditions such that the species replicates itself twice a week, equation 1 becomes

$$N_{t+1} = 3N_t \qquad (2)$$

(with a 3 instead of a 2, because before we had the individual and the single offspring it produced, now we have the individual and the two offspring it produced). Now, beginning with a single individual, we have, in subsequent weeks, 3, 9, 27, 81, 243, and so on.

We can use this model in a more general sense to describe the growth of a population for any number of offspring at all (not just 2 and 3 as above). That is, write,

$$N_{t+1} + RN_t \qquad (3)$$

where R can take on any value at all. R is frequently called the finite rate of population growth (or the discrete rate).

It may have escaped notice in the above examples, but either of the series of numbers could have been written with a much simpler mathematical notation. For example, the series 2, 4, 8, 16, 32, is a actually 2^1, $2^2, 2^3, 2^4, 2^5$, and the series 3, 9, 27, 81, 243, is actually $3^1, 3^2, 3^3, 3^4, 3^5$. So we could write,

$$N_t = R^t \tag{4}$$

which is just another way of representing the facts as described by equation 3. (Remember, we began with a single individual, so $N_0 = 1.0$.)

We now wish to represent the constant R (of equation 4) in a different fashion, to make further exposition easier. It is a general rule that any number can be written in many ways. For example, the number 4 could be written as $8/2$, or $9 - 5$, or 2^2, or many other ways. In a similar vein, an abstract number, say R, could be written in any number of ways: $R = 2b$, where b is equal to $R/2$, or $R = 2^b$, in which case $b = \ln(R/2)$ (where ln stands for natural logarithm). If we represent R as 2.7183^r, a powerful set of mathematical tools becomes immediately available. The number 2.7183 is Euler's constant, usually symbolized as e (actually 2.7183 is rounded off and thus only approximate). It has the important mathematical property that its natural logarithm is equal to 1.0.

So, rewrite equation 4 as,

$$N_t = e^{rt} \tag{5}$$

which is the classical form of the exponential equation (where R has been replaced with e^r). One more piece of mathematical manipulation is necessary to complete the toolbag necessary to model simple population growth. Another seemingly complicated but really rather simple relationship that is always learned (but frequently forgotten) in elementary calculus is that the rate of change of the log of any variable is equal to the derivative of that variable divided by the value of the variable. This rule is more compactly stated as,

$$\frac{d(\ln N)}{dt} = \frac{1}{N}\frac{dN}{dt} \tag{6}$$

So, if we rewrite equation 5 as,

$$\ln(N_t) = rt$$

we can differentiate with respect to t to obtain,

$$\frac{d(\ln N)}{dt} = r \tag{7}$$

and we can use equation 6 to substitute for the left-hand side of 7 to obtain,

$$\frac{dN}{Ndt} = r$$

and after multiplying both sides by N, we obtain,

$$\frac{dN}{dt} = rN \tag{8}$$

Equations 5 and 8 are the basic equations that formally describe an exponential process. Equation 8 is the differentiated form of equation 5, and equation 5 is the integrated form of equation 8. They are thus basically the same equation (and indeed are quite equivalent to the discrete form—equation 3). Depending on the use to which they are to be put, any of the above forms may be used, and in the ecological literature one finds all of them. Their basic graphical form is illustrated in figure 1.1.

In the examples of exponential growth introduced above, the parameter (r, or R) was introduced as a birth process only. The tacit assumption was made that there were no deaths in the population. In fact, all natural populations face mortality, and the parameter of the exponential equation is really a combination of birth and death rates. More precisely, if b is the birth rate (number of births per individual per time unit), and d is the death rate (number of deaths per individual per time unit), the parameter of the exponential equation is

$$r = b - d \tag{9}$$

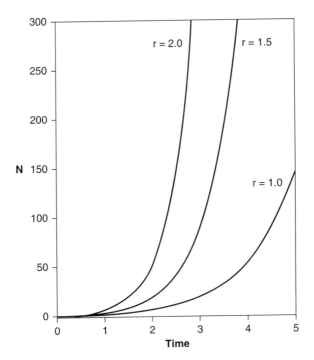

Figure 1.1. Graphs of equation 5.

where the parameter r is usually referred to as the intrinsic rate of natural increase.

One other simplification was incorporated into all of the above examples. We presumed always that the population in question was initiated with a single individual, which almost never happens in the real world. But the basic integrated form of the exponential equation is easily modified to relax this simplifying assumption. That is,

$$N_t = N_0 e^{rt} \tag{10}$$

which is the most common form of writing the exponential equation. Thus, there are effectively two parameters in the exponential equation: the initial number of individuals, N_0, and the intrinsic rate of natural increase, r.

Putting the exponential equation to use requires estimation of the two parameters. Consider, for example, the data presented in table 1.1.

TABLE 1.1
Number of Aphids Observed per Plant in a Corn Field

Date	Number of Aphids	ln(Number of Aphids)
March 25	0.02	−3.91
April 1	0.5	−0.69
April 8	1.5	0.40
April 15	5	1.61
April 22	14.5	2.67

Here we have a series of observations over a 5-week period of the average number of aphids on a corn plant in an imaginary corn field.

As a first approximate assumption, let us assume that this population originates from an initial cohort that arrived in the milpa on March 18 (one week before the initial sampling). We can apply equation 10 to these data most easily by taking logarithms of both sides, thus obtaining,

$$\ln(N_t) = \ln(N_0) + rt \tag{11}$$

which gives us a linear equation of the natural logarithm of the number of aphids versus time (where we code March 18 as time = 0, March 25 as time = 1, April 1 as time = 2, etc. . . . .). Figure 1.2 is a graph of this line along with the original data to which it was fit, and figure 1.3 is a graph of the original data along with the fitted curve on arithmetic axes.

From these data we estimate 1.547 aphids per aphid per week added to the population (i.e., the intrinsic rate of natural increase, r, is 1.547, which is the slope of the line in figure 1.2). The intercept of the regression is −4.626, which indicates that the initial population was 0.0098 (that is, the anti ln of −4.626 is 0.0098), which is an average of about one aphid per 10 plants. Now, if we presume that once the plants become infected with more than 40 aphids per plant the farmer must take some action to try and control them, we can use this model to predict when, approximately, this time will arrive. The regression equation is,

$$\ln(N \text{ of aphids}) = -4.626 + 1.547t$$

which can be rearranged as,

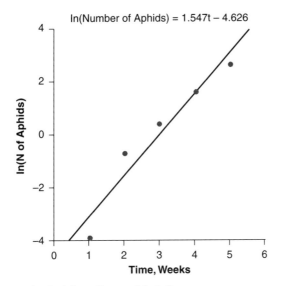

Figure 1.2. Plot of aphid data (from table 1.1).

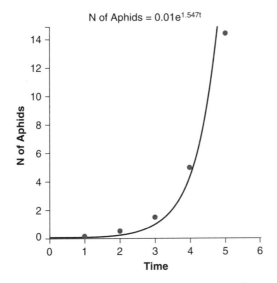

Figure 1.3. Aphid data from figure 1.2 plotted arithmetically.

$$t = [\ln(N \text{ of aphids}) + 4.626]/1.547$$

The natural log of 40 is 3.69, so we have,

$$t = (3.69 + 4.626)/1.547 = 5.375$$

Translating this number into the actual date (April 22 was time = 5), we see that the critical number will arrive about April 24 (actually at 3:00 P.M. on April 24, theoretically).

Naturally, the natural world contains many complicating factors, and the exact quantitative predictions made by the model could be quite inaccurate. As we discuss in later chapters, including some of the complicating factors will increase the precision of the predictions. On the other hand, April 24 really does represent the best prediction we have, based on available data. It may not be a very good prediction, but it is in fact the only one available. It may seem quite counterintuitive that, having taken five full weeks to arrive at only 14 aphids per plant, in only 2 more days the critical figure of 40 aphids per plant will be reached, but such is the nature of exponential processes. A simple model like this could help the farmer plan pest control strategies.

Density Dependence: Intraspecific Competition

In the above section, we showed that any population reproducing at a constant per capita rate will grow according to the exponential law. Indeed, that is the very essence of the exponential law; each individual reproduces at a constant rate. However, the air we breath and the water we drink are not completely packed with bacteria or fungi or insects; as they would be if populations grew exponentially forever. Something else must happen. That something else is usually referred to as intraspecific competition, which means that the performance of the individuals in the population depends on how many individuals are in it; this concept is commonly known as density dependence. Density dependence is a complicated issue, one that has inspired much debate and acrimony in the past and one that still forms an important base for more modern developments in ecology.

The idea of density dependence was originally associated with the human population and was brought to public attention as early as the eighteenth century by Sir Thomas Malthus (1830). Verhulst (1838) first formulated it mathematically as the "true law of population," better known today as the logistic equation (see below). Later, Pearl and Reed (1920), in attempting to project the human population size of the United States, independently derived the same equation. While this mathematical formulation was being developed, Gause (1934) and other biologists initiated a series of laboratory studies with microorganisms in which population growth was studied from the point of view of competition, both intra- and interspecific.

In the early part of the twentieth century a variety of terms appeared, all of which essentially referred to the same phenomenon: a population approached some sort of carrying capacity through a differential response of per capita population growth rate to different densities. Chapman (1928) formulated the idea in terms of "environmental resistance." In 1928, Thompson redefined Chapman's formulation as "general" and "independent" of density versus "individualized" and "dependent" on density, and later Smith (1935) proposed the density-independent–density-dependent gradient. Thus, by the 1930s the dichotomy of density independence versus density dependence had taken firm root, after having been sown not long after the turn of the twentieth century.

In the 1930s Nicholson and Bailey (1935) first formalized the concept of regulation through density-dependent factors and clearly identified the idea of intraspecific competition with density dependence (see also Nicholson 1957). In Nicholson and Bailey's conceptualization of density dependence, four points were proposed: (1) population regulation must be density dependent; (2) predators and parasites may function as density-dependent forces; (3) more than density dependence alone may function to regulate population density; and (4) density dependence did not always function to regulate population density.

In contrast to Nicholson and Bailey (and especially their later followers), Andrewartha and Birch (1954) held that the environment was not divisible into density-dependent versus density-independent forces. Andrewartha and Birch argued that, although resources could limit populations, they rarely do so because some aspect of the physical environment (usually collectively referred to as the weather) almost always reduces the population before it becomes stressed by lowered resources. They furthermore noted that the mathematical models that presume equilibrium and persistence were not really necessary if there was no "balance" in nature (density dependence strongly implies some sort of balance of nature). Most data sets failed to support the idea of density dependence, and Andrewartha and Birch suggested that the idea was possibly untestable. Rather, they argued, the regulation of populations was frequently taken as an article of faith. The problem was, How long could a population persist without regulation? Their recognition of the fact that local populations would frequently go extinct but would be refounded from other population centers anticipated ideas of metapopulations that would become popular some 20 years later.

Milne (1961) modified both versions of population regulation (the density-dependent and density-independent schools) and noted that perfect density dependence, if it ever exists, does so only at very high densities. Rather, what most characterizes populations in nature is what might be referred to as imperfect density dependence (similar to what Strong [1986] referred to as "density vagueness"); predators and parasites plus density-independent effects usually held populations below levels at which intraspecific competition could become important.

Finally, several variations on the basic theme have recently emerged. Levins (1974) introduced the notion of positive and negative feedback loops for analyzing the dynamics of a population. Dempster (1983) suggested that density independence could be operative within limits, such that an upper ceiling would be imposed upon the population and a lower limit would prevent the population from going extinct. Almost all of these variations are fundamentally in the density-dependence camp but with strong notions of nonlinearities and the importance of spatial distribution, topics discussed in later chapters.

In the end, it would seem that the entire debate about density-dependent versus density-independent control of populations was focused on a false dichotomy. In a variety of guises (e.g., metapopulations, as discussed in chapter 5), modern ecology has come to acknowledge that density-dependent and density-independent forces may function together to regulate populations in nature. But more important, there is general agreement that the rate of growth or decline of a population relative to its size does not necessarily suggest any particular mechanism of regulation. However, there is a recent burgeoning literature, beyond the intended scope of this text, that seeks to use advanced methods of analysis of long-term data sets to determine whether density dependence operates (Hastings et al. 1993). Part of this later literature is associated with the possibility that many populations under density-dependent control actually may be chaotic (discussed more fully in chapter 3). Chaotic populations can easily be confused with random populations. One way of resolving some of the earlier debates about density dependence was to acknowledge that extreme density dependence (which would promote chaos) could easily produce population behavior that looked quite density independent (i.e., chaotic) (Gukenheimer et al. 1977).

As one can see from the previous paragraphs, the literature on density dependence is enormous. Yet much of it can be divided conceptually into three categories. First, the effect of density on the growth rate of the population (be it through declining reproduction or increased mortality) is simply added to the exponential equation to form the famous logistic equation (as discussed below). Traditionally, the logistic equation is expressed in continuous time as a differential equation, but recently a large literature has been generated by consideration of the special properties of the logistic idea expressed in discrete time, the logistic map. The logistic equation, either its continuous or discrete form, treats the population growth rate as a single constant, even though we understand it actually represents birth rate minus death rate. Other approaches treat each of these rates separately.

Decomposing the population growth rate into its two components, the second category focuses on the relationship between density and reproduction (i.e., density modifies birth rate). We guess that the first acknowledgment of density dependence in nature was by the world's earliest farmers. When one is planting crops, it soon becomes apparent that higher planting densities provide higher yields (which, in principle, are correlated with reproductive output), but only to a point. Once you reach a high enough density, further increases in density fail to provide further increases in yields. This general relationship is referred to as the yield–density relationship and is, in some respects, the most elementary form of density dependence. Originally developed mainly in the agronomy literature, the relationship between density and yield subsequently became an important theoretical baseline for general plant ecology. Yield was usually a product of reproduction, because the subject was mostly the yield of seed crops such as corn and soybeans, and thus the subject of yield and density can be thought of more generally as the relationship between density and reproduction.

Finally, the third category examines the possibility that density affects survivorship rather than reproduction (i.e., density modifies death rate). The main literature on this topic was originally conceived by forest ecologists, but the idea has since been generalized as self-thinning laws, which are mainly used in plant ecology. The yield–density relationship, discussed in the previous paragraph, involves examining yields of different populations that have been sown at different densities. It is a static approach in this sense. Once established through sowing, the population density remains constant and the variable of interest is the yield. An alternative approach is more dynamic and follows changes in both size (biomass) and density over time in the same population. This more dynamic approach considers mortality as well as growth, and in the context of forestry, where it was originally developed, mortality is known as thinning.

In the following three sections, we follow this basic schema: (1) the logistic equation, (2) yield–density relations, and (3) self-thinning laws.

The Logistic Equation

Density dependence is generally regarded as the major modifier of the exponential process in populations. Consider the data shown in figure 1.4 for example (Vandermeer 1969). The protozoan *Paramecium bursaria* was grown in bacterial culture in a test tube, and the data shown are for the first 11 days of culture (data are number of cells per 0.5 ml). In figure 1.5 those numbers are shown as a graph of ln$N$ versus time (recall how the intrinsic rate of natural increase was estimated in this way).

The relationship is approximately linear (see figure 1.5), and our conclusion would be that the population is growing according to an exponential law. If this equation were followed into the future, we would have a very large population of *Paramecium*. Indeed, considering the size of *P. bursaria*, there could be about 3000 individuals in 0.5 ml if you stacked them like sardines. Thus, the 3001st individual would cause all the animals to be squeezed to death, and we can compute exactly when this event will happen.

$$\ln(3001) = 0.337t + 1.239$$

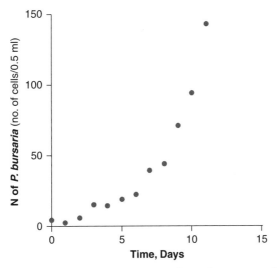

Figure 1.4. Growth of a culture of *Paramecium bursaria* in a test tube (Vandermeer 1969).

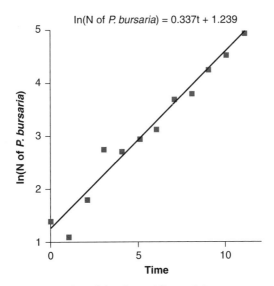

Figure 1.5. Logarithmic plot of the data of figure 1.4.

which can be arranged to read,

$$t = [\ln(3001) - 1.239]/0.337 = 20.80$$

Thus, on the basis of an 11-day experiment, we can conclude that after about 20 days, the test tube will be jam-packed with *P. bursaria* such that all the individuals will suddenly die when that 3001st individual is produced. The actual data for the experiment carried out beyond the 20-day expected protozoan Armageddon is shown in figure 1.6. These data suggest that something else happened. As the density of the *Paramecium* increased, the rate of increase declined, and eventually the number of *Paramecium* reached a relatively constant number. The theory of exponential growth must be modified to correspond to such real-world data.

Begin with the exponential equation, but assume that the intrinsic rate of growth is directly proportional to how much resource is available in the environment. Thus, we have,

$$dN/dt = rN \qquad (12)$$

the classical exponential equation discussed earlier in this chapter. But here we presume that r is directly proportional to F ($r = bF$), where F is the amount of resource (F for food) in the system that is available for the population and the constant b represents the efficiency of converting food to babies. Thus, equation 12 becomes,

$$dN/dt = bFN \qquad (13)$$

But now we assume that there is no inflow of resource into the system so that the total amount of resource is constant and is divided up into that part that is useable by the population and that part that has already been used. That is,

$$F_T = F + cN \qquad (14)$$

where F_T is the total amount of resource in the system and c is the amount of F held within each individual in the population. Equation 14 can be manipulated to read,

$$F = F_T - cN \qquad (15)$$

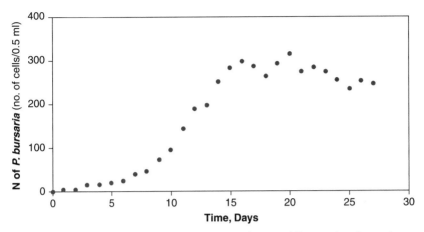

Figure 1.6. Long-term data for the same population of *Paramecium bursaria* (data from time $=0$ to time $=12$ are the same as in figure 1.4).

Substituting equation 15 into equation 13, we have,

$$dN/dt = b(F_T - cN)N \qquad (16)$$

whence we see that equation 16 is a quadratic equation. Finding the equilibrium point, that is, the point at which the population neither increases nor decreases, is done by setting the population growth rate equal to zero, thus obtaining,

$$0 = b(F_T - cN)N$$

which has two solutions. The first solution is at $N=0$, which simply says the rate of change of the population is zero when there are no individuals in the population. The second solution is at F_T/c, which is the maximum value that N can have. This is the value of N for which $F=0$, when all the resource in the system is contained within the bodies of the individuals in the population. Because the limitations of the environment are more or less stipulated by the value of F_T, and the maximum number of individuals that that environment can contain is F_T/c, the value F_T/c is frequently referred to as the carrying capacity of the environment (the capacity the environment has for carrying individuals). The traditional symbol to use for carrying capacity is K, so we write $K = F_T/c$. We also note that as the population approaches zero (as N becomes very small but not exactly at zero), the rate of increase of the original exponential equation will be bF_T (since the general equation is bF and when N is near to zero F is almost the same as F_T). After some manipulation of equation 16 we can write,

$$dN/dt = bF_T N[(F_T/c) - N)/(F_T/c)]$$

and now substituting $r = bF_T$ and $K = F_T/c$ we obtain

$$\frac{dN}{dt} = rN \frac{(K-N)}{K} \tag{17}$$

which is the classic form of the logistic equation. Note the form of the equation. It has a very simple biological interpretation. The quantity $(K-N)/K$ is the fraction of total available resource that remains available: that is, the fraction of the carrying capacity that has not yet been taken up by the individuals in the population. As shorthand we might refer to this quantity—the fraction of the carrying capacity or the fraction of total available resource—as the available niche space. Then the logistic equation is obtained by multiplying the original intrinsic rate of increase, r, by the available niche space.

Returning to the earlier example of *Paramecium bursaria*, a glance at the data suggests that the carrying capacity is around 290 individuals (average all the points after the data have leveled off). The original estimate of r as 0.337 was probably too low (since the effects of density dependence were probably effective even during the time of the initial growth) so taking a slightly larger value, let $r = 0.5$. The logistic equation for these data then becomes,

$$\frac{dN}{dt} = 0.5N \left(\frac{290 - N}{290} \right)$$

which is plotted in figure 1.7, along with the original data. This example represents a reasonably good fit to the logistic equation.

The existence of density dependence also calls into question the extrapolations that one is tempted to make from a process that seems inexorably exponential. The example earlier in this chapter of the aphids in the milpa agroecosystem is a case in point. Concluding that the

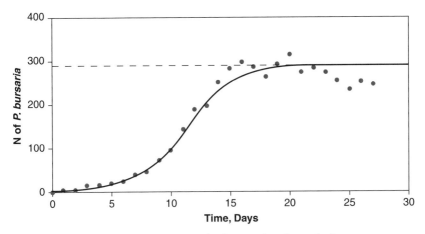

Figure 1.7. Fit of logistic equation to the *Paramecium bursaria* data.

farmer had only 2 days before disaster struck may have been correct, but it also could have been grossly in error, depending on the strength of the density dependence. Indeed, with strong density dependence, the field's carrying capacity for the herbivore could have been below the threshold where the farmer needed to take action, in which case no action at all would have been necessary.

In some management applications (e.g., fisheries), it is desirable to maximize the production of a population, which is to say maximize the rate of increase, not the actual population. The logistic equation can provide a useful guideline for such a goal because it is reasonably simple to show that the maximum rate of increase of the population will occur when the population is equal to $K/2$. Thus, once the carrying capacity is known, the population density at which the rate of growth will be maximized is automatically known. In actual practice this so-called maximum sustained yield has some severe problems associated with it, largely stemming from the simplifying assumptions that go into its formulation (these issues are more fully discussed in chapter 4).

Using much of the same reasoning as above, we can formulate density dependence in discrete time rather than continuous time. Rather than asking how a population grows instantaneously, we can ask how many individuals will be in the population next year (or some other time unit in the future) as a function of how many are here now. Recall equation 3 from the section on the exponential equation,

$$N_{t+1} = RN_t$$

which is a statement of population growth in discrete time. Now, rather than proceed with a generalization about what numbers will be in a future time (which was the development taken earlier), we remain in the realm of discrete time and ask what might be the necessary modifications to make this equation density dependent. In other words, what do we come up with if we use the same rationale we used in developing the logistic equation, but this time do it in discrete time?

It seems reasonable to suppose that the population will grow slowly if the population is near its carrying capacity (K) and will grow more rapidly if it is far below its carrying capacity. This is the same as saying that R varies with population density. If we simply allow R to vary with density (the same conceptual approach we took with the logistic equation), we could write,

$$R = r(K - N_t)/K$$

which would make the original equation

$$N_{t+1} = r[(K - N_t)/K]N_t$$

Frequently, the carrying capacity is set equal to 1.0, a transformation that does not change the qualitative behavior of the equation and makes it easier to work with. Thus we have,

$$N_{t+1} = rN_t(1 - N_t)$$

(Note that the parameter r here refers to discrete population growth, whereas earlier it refers to continuous growth.) This equation is usually referred to as the logistic map (map, because it maps N_1 into N_2) or the logistic difference equation. It has some remarkable features that will be explored in more detail in chapter 4. We add a small technical note here. The logistic map is not what you get when you integrate the logistic differential equation and then solve for N_{t+1} in terms of N_t, although the perceptive reader might be excused for thinking it so since both equations are called logistic. The logistic map is derived directly from first principles (as above). Integrating the logistic differential equation gives a different time interval map.

The Yield–Density Relationship

The process of intraspecific competition (density dependence) is certainly ubiquitous and thus legitimately calls for a theoretical framework, the most common and general of which is the logistic equation. However, for many applications it is not sufficient to consider only population growth rate. We also want to decompose that rate into its component parts, birth rate and death rate. In this section we consider the effect of density on birth rate. This theory developed from work on plants, especially in agroecosystems. Farmers need to know the relationship between planting density and the yield of a crop (which is frequently the seed output). This relationship is known as the yield–density relationship and is the basis of much agronomic planning as well as a springboard for much general plant ecology. For our purposes here, the yield–density relationship provides the most elementary form of the effect of intraspecific competition on reproduction and lays bare its essential elements. We thus give considerable space to the development of the principles of intraspecific competition as reflected in the yield–density relationship (Vandermeer 1984).

The formal elaboration of yield–density relationships first appeared in 1956 with Shinozaki and Kira's work. Shinosaki and Kira noted, as had many workers before them, that plotting yield versus density for various plant species usually results in a characteristic form. Several examples are shown in figure 1.8.

Shinozaki and Kira suggested a simple hyperbolic form:

$$Y = Dw_{max}/(1 + aD)$$

where D is population density, Y is yield, w_{max} is the unencumbered (i.e., without competitive effects) yield of an individual plant, and a is an arbitrary constant. This equation asymptotes as D becomes very large and thus corresponds to another well-known empirical observation in plant ecology known as the law of constant final yield (which actually is not always true as discussed below). Figure 1.9 shows Shinozaki and Kira's equation in comparison with the rape data of figure 1.8.

Much of this empirical curve fitting can be rationalized with some simple plant competition theory. We begin by considering what might happen with individual plants and later accumulate those plants into a

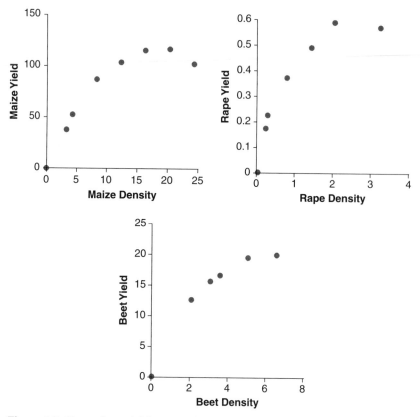

Figure 1.8. Exemplary yield versus density data (from Willey and Heath 1969).

population so as to examine the effect of density. Consider a single corn plant in a pot. When provided with all necessary light, water, and nutrients, it will grow to some specified height with some specified biomass. If two corn plants are planted in a pot of the same size and provided with the same amount of light, water, and nutrients, each of the corn plants will attain a biomass smaller than the corn plant grown alone, because the same amount of resources is being used by two individuals rather than one. If we symbolize the biomass attained when a plant is growing alone as k, we can write the simple relationship,

$$w_1 = k - \alpha w_2 \tag{18}$$

where w_1 refers to the biomass (usually estimated as dry weight) of the first plant, w_2 to the biomass of the second plant, and α is the propor-

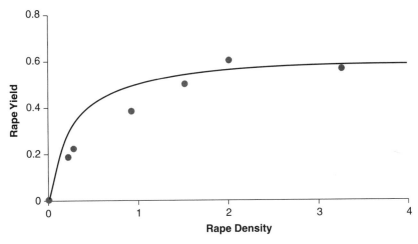

Figure 1.9. Equation of Shinozaki and Kira superimposed on the rape data from figure 1.8.

tionality constant that relates the decrease in biomass of the first individual as a proportion of the biomass of the second individual. Rearranging equation 18, we see that

$$\alpha = (k - w_1)/w_2 \qquad (19)$$

This same development could be applied to three plants growing in a single pot, in which case the equation describing the results would be,

$$w_1 = k - \alpha_{1,2} w_2 - \alpha_{1,3} w_3 \qquad (20)$$

where $\alpha_{1,2}$ is the effect of a unit of biomass of individual 2 on the biomass of individual 1 and $\alpha_{1,3}$ is the effect of a unit of biomass of individual 3 on the biomass of individual 1. The parameter α is frequently referred to as a competition coefficient because it represents the effect of one individual on another. The calculation of α from real data is quite easy when we have only two plants: Grow a single plant in a pot and measure its biomass after some specified time, giving the value of k; then grow two plants in a pot and measure their biomasses, giving the values of w_1 and w_2; then apply equation 18 to determine the value of α. The estimation of the competition coefficients when there are more than two individuals is somewhat more complicated but need not concern us at this point. For now it is important only to understand the logic of the thinking that went into the construction of equation 19. We now proceed to generalize equation 20.

Let us suppose that instead of just two or three individuals planted in a pot, we plant a large number of individuals. If the total number planted is n, we can expand equation 19 by simply adding more terms until we have added all n individuals to the calculation. That is, equation 20 for n individuals becomes,

$$w_1 = k - \alpha_{1,2}w_2 - \alpha_{1,3}w_3 - \ldots - \alpha_{1,n}w_n$$

or, more compactly,

$$w_1 = k - \Sigma\, \alpha_{i,j}w_j \tag{21}$$

where the summation is taken from $j=2$ to $j=n$. If all the individuals are exactly the same, it might be argued that all the $\alpha_{i,j}$ values are equal. As a first approximation this is probably a good assumption. However, there is a crucial way in which the competition coefficients differ from one another, as becomes evident when we try to elaborate this same example from the level of a pot to the level of a field, below.

For now assume (a bit unrealistically) that all individuals produce the same biomass and the competition between any two pairs of individuals is identical from pair to pair (or assume we can substitute the mean values for biomass and competition). The summation over i and j now represents the summation of two constants exactly D (density of the population) times, so we can thus write,

$$w' = k - D\alpha'w' \tag{22}$$

where D is the population density, and the primes in this case indicate mean values. Equation 22 can be rearranged as follows:

$$w' + D\alpha'w' = k$$

or

$$w'\,(1 + \alpha'D) = k$$

and finally,

$$w' = \frac{k}{(1+\alpha'D)} \qquad (23)$$

If w′ is the biomass of an average individual in the population, the total population yield must be,

$$Y = w'D$$

and substituting from equation 23 for w′, we obtain,

$$Y = \frac{Dk}{(1+\alpha D)} \qquad (24)$$

which is identical to the empirical equation of Shinozaki and Kira (we have eliminated the prime from the competition coefficient to make the notation less messy). The advantage of equation 24 is that, because the derivation is based on plant competition theory, the parameters in the equation have obvious meaning, k being the unencumbered yield of an average individual plant and α being the mean competition coefficient between two individual plants.

An additional complication arises when we have data like the maize data of figure 1.8, where at high densities the yield actually falls. To accommodate data such as these, Bleasdale and Nelder (1960) suggested modifying the basic Shinozaki and Kira equation with an exponent, citing either,

$$Y = \frac{kD}{(1+\alpha D)^b}$$

or

$$Y = \frac{kD}{(1+\alpha D^b)}$$

as reasonable approximations to data that are shaped parabolically. The constant b is, in the context of Bleasdale and Nelder's derivation, a fitted constant that they presume is related to an allometric effect (i.e., the harvested material is produced proportionally less at higher densities of the plant). Either equation reduces to Shinzaki and Kira's equation when $b = 1.0$. Bleasdale and Nelder chose the first of those two equations arbitrarily, and it has become a standard in the plant ecology literature. It is worth noting that it is not only the allometric effect that can produce a yield density curve that descends at high densities. An increase in competitive intensity as individuals get closer to one another also will create the effect of declining yield with high density (Vandermeer 1984).

Either of Bleasdale and Nelder's equations can be viewed as a discrete map, much like the logistic map, although with slightly different properties. If we think of yield as the number of organisms that will be found in the population in the next generation, this equation becomes equivalent to an iterative map (like the logistic map). The properties of these sorts of maps will be discussed in detail in chapter 4.

Density Dependence and Mortality: Thinning Laws

In the above developments, we assumed that density dependence acts in such a way that the growth of individuals is slowed by a larger population and that a decline in individual growth rate leads to a lower birth rate that eventually stabilizes the population at some particular number. In our development of the logistic equation, no explicit assumption was made about birth or mortality, and the derivation revolved around the intrinsic rate of natural increase, which includes both death and birth rates. However, implicitly in the section on the logistic and explicitly in the above section on the yield–density relationship, the assumption was that we were dealing exclusively or mainly with birth rate modifications rather than death rate modifications.

There are times where the distinction between birth rate and death rate modifications can be crucially important. For example, the growth in biomass of a plantation of trees is usually approximately logistic in form, but the same logistic equation could account for the pattern in either figure 1.10A or 1.10B. And the difference between the two figures is not trivial from a forester's point of view. In figure 1.10A there are large numbers of very small trees, none of which is harvestable; in figure 1.10B there are a smaller number of larger trees. The point is that in figure 1.10A there has been a great deal of intraspecific competition, but it took the form of each individual's growing slowly and almost no mortality, whereas in figure 1.10B one of the main responses to in-

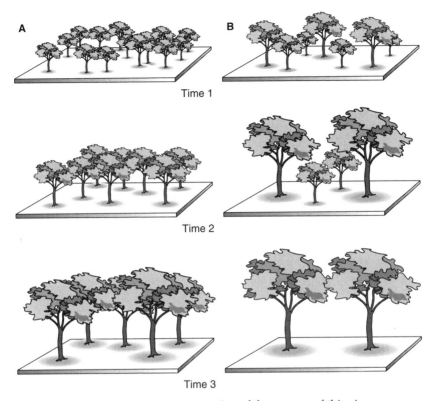

Figure 1.10. Diagrammatic representation of the process of thinning.

traspecific competition was for some individuals to die while others continued growing rapidly. The biomass of the forests in the two figures is the same (that is the way the example was constructed), but one will be useful for harvest, the other not. Similar examples could be given for any organism with indeterminate growth. For example, many fish become stunted when in very dense populations and thus represent less of an attraction for sport or commercial fishing.

Reflect for a moment on the pattern of growth and mortality in a densely planted tree plantation or in a natural forest when large numbers of seeds germinate more or less simultaneously. First, seedlings are established at a very high density. Walking through a beech–maple forest, for example, one is struck by the carpet of maple seedlings in almost every light gap one encounters. As the seedlings grow, the increase in biomass of each individual treelet is limited by intraspecific competition. But when populations are sufficiently dense, inevitably some individuals come to "dominate" (grow larger) while others become "suppressed" (remain small owing to competition from their neighbors). Eventually, the suppressed individuals die, and we say that the population has been thinned. But then the trees keep on growing and the process repeats itself; some trees are suppressed, others dominate. In this way, a population of plants that began at a very high density is thinned to the point that the adults are at some sort of carrying capacity. In some ways, this process seems to be the reverse of what was described in the development of the logistic equation. Here, we begin with a number larger than K, and through the process of thinning the population is reduced to K, rather than beginning with a small population and increasing to the value of K. On the other hand, remember that biomass is increasing throughout the process.

This phenomenon is most easily seen as a graph of log of biomass versus log of density at harvest time, as shown with the data in figure 1.11 and more schematically in figure 1.12. But the whole idea is much more dynamic. To look at the dynamics, take a single starting density and observe changes in biomass (or some related variable). If no mortality occurs, we expect a straight vertical line; that is, the per plant biomass increases, but the population density remains constant (see figure 1.12A). But if there is mortality, over time, the curve will shift to the left, to lower densities, while at the same time the per plant biomass will have increased (see figure 1.12B). If, on the other hand, we had begun with two different populations at slightly different densities we would see that plants of both populations would increase biomass. Assuming that densities were such that this increase in biomass happened without competition, the two populations will grow in biomass the same amount (see figure 1.12C). Now if we let both of these populations continue to grow, we expect some thinning (mortality) to occur, especially in the denser population (figure 1.12D).

Here, we can see both a plastic effect on growth and a mortality effect. The plastic effect is a smaller biomass increase at larger densities, as shown in figure 1.12D. The mortality effect is seen as a decrease in density at higher densities, as shown in figure 1.12D. If we continue the pattern of development illustrated in figure 1.12 through time, we see that each population begins its process of thinning as it approaches a theoretical thinning line, as can be seen in the data shown in figure 1.13. Once mortality starts, the population tends to follow a straight

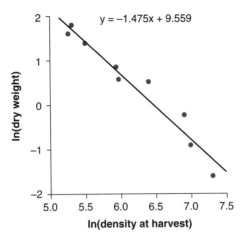

$$y = -1.475x + 9.559$$

Figure 1.11. Typical relationship between log density of surviving plants and natural log (ln) dry weight per plant. Example is of *Helianthus annuus* (Hiroi and Monsi 1966).

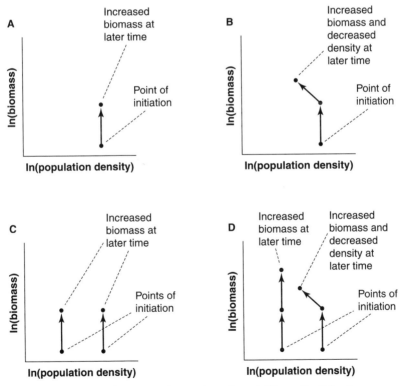

Figure 1.12. Expected pattern of growth and mortality as the thinning process and growth in biomass interact. Log of biomass refers to the logarithm of the biomass of an average individual in the population.

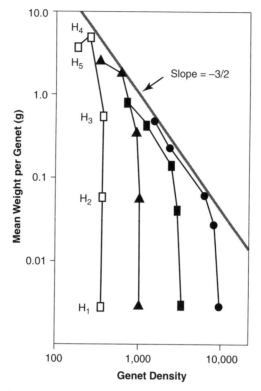

Figure 1.13. Relationship between the density of genets and the mean weight per genet in populations of *Lolium perenne*. H_1, H_2 etc. are successive harvests (Kays and Harper 1974).

line on a log-log scale. This kind of relationship has been shown many times—most often in herbaceous plants over time or in comparisons of woody plant plantations at different densities.

This self-thinning law (called *self* because no forester or agronomist is there doing it) was first developed for plants but many animals show a similar pattern. This is a very nice way of showing growth and mortality effects of density on the same graph. But it also provides an elegant way of looking at density-dependent mortality that can be easily compared among species on very different time scales because time is not explicit. Furthermore, some time ago plant ecologists noticed that this process of self-thinning always seemed to take on a particular pattern. In plots of the logarithm of the biomass of an average individual plant versus the population density at the time the biomass was measured, the points in a thinning population appeared linear, and the slope of the line always appeared to be nearly $-3/2$ (as in figures 1.11 and 1.13), providing the population was undergoing thinning. This phenomenon is known as the three-halves thinning law.

Yoda and colleagues (1963) provided an elegant theory explaining the origin of the law. Suppose that each plant is a cube. If each side of the cube is x, the area of one of the cube's faces is x and the volume of the cube is x^3. Now we imagine that the plantation is made up of a large number of these cubes and they begin growing and thinning through intraspecific competition. The overall process is illustrated in figure 1.14. The area of the plantation is A. The population density will be the total area divided by the surface area occupied by a single plant (that is, a single cube). Thus, D, the population density, is equal to A/x^2. We now presume that the biomass, w, of an individual plant is approximately equal to the volume of the cube representing it, so that $w = x^3$. So we have the pair of equations,

$$D = A/x^2$$

and

$$w = x^3$$

Rearranging these equations we write,

$$x = A^{1/2}D^{-1/2}$$

and

$$x = w^{1/3}$$

Since the left-hand side of both equations is equal to x, we can set the right-hand sides of the two equal to each other, giving,

$$A^{1/2}D^{-1/2} = w^{1/3}$$

which simplifies to,

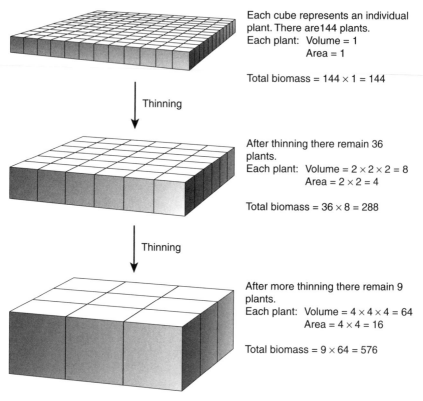

Each cube represents an individual plant. There are 144 plants.
Each plant: Volume = 1
            Area = 1

Total biomass = 144 × 1 = 144

Thinning

After thinning there remain 36 plants.
Each plant: Volume = 2 × 2 × 2 = 8
            Area = 2 × 2 = 4

Total biomass = 36 × 8 = 288

Thinning

After more thinning there remain 9 plants.
Each plant: Volume = 4 × 4 × 4 = 64
            Area = 4 × 4 = 16

Total biomass = 9 × 64 = 576

Figure 1.14. Yoda and colleagues's interpretation of the origin of the 3/2 thinning law.

$$w = A^{3/2}D^{-3/2}$$

which can be put in the more standard form,

$$\ln(w) = (3/2)\ln(A) - (3/2)\ln(D)$$

which represents a straight line with a slope of $-3/2$ on a graph of $\ln(w)$ versus $\ln(D)$.

Thus we see from very simple reasoning that it is not unusual to expect the three-halves thinning law. On the other hand, the basic empirical base of the "law" has been persistently questioned (e.g., Westoby 1984). In fact only a few data sets conform convincingly exactly to a $-3/2$ slope, and many ecologists feel that data such as those shown in figures 1.12 and 1.13 are actually exceptions to a rule that is something other than −3/2. It is worth noting that, alongside this general consensus that the three-halves thinning law is not correct, its theoretical basis is quite shaky to start with. Plants are not, after all, cubes, and the fact that their thinning pattern does not follow the 3/2 thinning law exactly is not all that surprising. But the basic idea still seems sound. We should expect a linear relationship between the log of the biomass and the log of the density, and that appears to be almost always true. While we expect the slope of that line to be −3/2 in the case of plants shaped like cubes, most plants are not shaped like that. It is perhaps best to treat the 3/2 thinning law something like the Hardy-Weinberg law (Futuyma 1979), something that ought to be true under ideal conditions but rarely happens in fact because those ideal conditions are hardly ever met.

2
C H A P T E R

Life History Analysis

In chapter 1 we saw how simple mathematical models can illustrate general principles of population dynamics. Such models can be used to examine the range of dynamics we might observe in real populations over the long run and to make at least short-term population projections.

Population dynamic models are also useful for understanding a very different problem: life history evolution. Using such models, we can examine the population consequences of evolutionary changes in particular parameters such as age or size at first or last reproduction; age or stage-specific birth, growth, and death rates; size at birth; number and size of offspring; longevity, and so on. The study of life history has a long history in ecology, from the classic studies of Lack (e.g., 1947) and Cole (1954) to the present time. Cole (1954), for example, showed that adding one offspring in a previous year could have much bigger effects on long-term population growth than adding several more offspring in a current year. Using population models has become standard in this field. If, for example, evolutionary pressure is exerted on individuals within a population to have fewer offspring, modeling that population with the logistic equation would begin with the assumption that evolutionary pressure was causing the intrinsic rate of natural increase to decrease. On the other hand, if evolutionary pressure on the individuals in a population caused each individual to require more space in the environment, we might model that situation with a logistic equation again, but this time incorporating a decreasing K. In either case we can ask what will happen to the population over the long run if these evolutionary pressures are brought to bear. This is the central problem of life history evolution: How do life history traits evolve?

The rhinoceros mother cares for her young far more closely than most other species do, a life history style evolved under certain conditions.

The problem is usually approached by asking how proposed patterns of selection pressure (which inevitably act on individuals) will affect the population as a whole.

Life history traits are traits that affect basic survival and reproductive schedules of organisms (e.g., age-specific birth rate, stage-specific growth rate, crude death rate, size at birth, number of offspring, size of offspring, longevity). The key assumption that underlies life history theory is that there are tradeoffs among these life history components. If there were not such tradeoffs, it is fairly obvious what natural selection would ultimately produce—immortal individuals who start reproducing at birth and continue producing many large offspring in perpetuity. This is obviously not what happens and the reason is that there are costs associated with reproduction. High reproduction is negatively correlated with other components of fitness. For example, high investment in reproduction early in life likely reduces subsequent reproductive success. Similarly, producing large offspring usually means that fewer of them can be produced.

In this chapter, we concentrate on the most important life history tradeoff: the relationship between investment in survivorship and in reproduction. First, we describe the classic life history ideas of r versus K selection, which come directly out of the logistic equation. Although these ideas imply a tradeoff between survivorship and fecundity, this is not explicit, because birth and death rates are not parameters in this equation. Second, we ask how to quantify the cost of reproduction that is the basis for this tradeoff: high current fecundity should mean lower survival (because organisms are forced to allocate fewer resources to maintenance, growth, and defense) or lower subsequent reproduction, or both. The assumption of a cost of reproduction is fundamental to life history theory, but testing it is not simple. Third, assuming that there is a cost to reproduction, we ask, What is the optimal schedule of reproduction (when and how much to allocate to reproduction) and, consequently, survival.

Investment in Survivorship versus Reproduction: The r, K Continuum

Life history evolution theory can be traced to the early work of Cole and Lack, but its birth in modern form emerged in musings of Cody

(1966) and MacArthur and Wilson (1967) based on the logistic equation described in chapter 1. The basic idea is that, at low density, we expect per capita population growth rate to be primarily influenced by the value of r, the intrinsic rate of natural increase. ($[K-N]/K$ almost equals $K/K = 1$, so dN/Ndt approximates r). However, at high density, per capita population growth rate will be primarily influenced by the value of K, the carrying capacity. The fact that the per capita population growth rate is a reflection of individual fitness suggests that natural selection will act primarily on traits that increase r at low density and on traits that influence K at high density. Hence the appellation r-K selection. Although the formal basis of r-K selection theory is strictly that of density-dependent selection (Boyce 1984), the idea rapidly broadened to consider any environmental conditions that might lead to low versus high density. Thus, r-selection was assumed to be the dominant mode of selection early in succession, in highly disturbed environments, and in unpredictable environments, whereas K-selection was assumed to be the dominant mode of selection in more stable and predictable conditions (table 2.1). Ecologists then began to apply r and K selection to life history traits, particularly the tendency of organisms to take one or another extreme approach to investing in reproduction and survival (Cody 1966, Pianka 1970). Some species tend to produce large numbers of offspring, begin reproduction early, and die young. This strategy is associated with r selection. Other species tend to produce small numbers of offspring but invest considerably in parental care, begin reproduction later in life, and live relatively long lives. This strategy is associated with K selection. The attributes associated with each end of this continuum are listed in table 2.1. The idea in its most general form is that we can imagine populations existing in an environmental continuum ranging from unpredictable or harsh to constant or mild. Such terminology is vague enough to encourage some analysts to fit any observation whatsoever into the scheme; it is also vague enough to cause others to criticize its utility in the first place (e.g., Hairston et al. 1970, Wilbur et al. 1974).

Another major critique of the r-K paradigm is that the traits typically associated with K-selection (table 2.1) have little to do directly with the population carrying capacity. Traits that would enable more individuals to persist with a given amount of resource do not necessarily stem from or translate into larger or smaller K. That is, K is not a demographic parameter and so does not fit easily into a theory of evolution

TABLE 2.1
Some Attributes of r- and K-Selected Species and the Environmental
Characteristics That Select for Them (Pianka 1970)

	r-selection	K-selection
Mortality	Variable and unpredictable	More constant and predictable
Population Size	Variable, below carrying capacity	Constant, close to carrying capacity
Intra- and Interspecific Competition	Variable, often weak	Usually strong
Selection Favors	Rapid development Early reproduction Small body size Semelparity	Slow development Delayed reproduction Large body size Iteroparity
Length of Life	Usually shorter	Usually longer
Leads to	High productivity	High efficiency

of life history traits. Critics have thus argued that characteristics upon which selection acts are not really related to these particular phenomenological categories (i.e., intrinsic rate of natural increase and carrying capacity) but rather to characteristics associated with birth rates and death rates (Hairston et al. 1970, Wilbur et al. 1974). Indeed, most recent work on life history evolution (e.g., Roff 1992, Stearns 1992, Silvertown and Dodd 1996) emphasizes birth and death rates (i.e., characteristics associated with them), even though the r and K selection paradigm remains a useful heuristic tool for many workers.

To go beyond the simple ideas of r-K selection, we must deal with models that are more complicated than the simple logistic because natural selection operates mainly at the individual level and the interesting questions have to do with how that natural selection operates differently on different individuals. If all individuals are the same, as we presume in the exponential and logistic models, there is not much to ask beyond the simple questions about r and K as outlined above. Thus, the rest of this chapter and, indeed most of the rest of this text, is an explanation of these more complicated models. We can no longer assume, as we did in chapter 1, that all individuals in a population are identical. Few populations in nature would meet this assumption in any serious way.

To understand the patterns of evolution of traits that affect demography is the central goal of the study of life history evolution. The general strategy in the post-r-K world has been to focus on birth and death schedules and assume that natural selection maximizes the per capita growth rate of the population. It may be worthwhile to consider the utility of this underlying assumption in the first place, but here we are concerned with reviewing some of what has been done, not with criticizing it.

The main conceptual framework is that of tradeoffs. We begin by assuming that no organism can do everything. On the one hand, there are phylogenetic constraints on natural selection (which is why there are no six-legged vertebrates or four-legged insects). On the other hand, there are tradeoffs that limit the potential of natural selection. There are costs of reproduction and survivorship such that perhaps high investment in reproduction early in life reduces subsequent reproductive success or making large offspring means that fewer can be made. So, we have to ask what the optimal reproductive and survivorship schedules are, given the relevant costs, and how those optima change as the costs change.

The rest of this chapter is organized around two general questions:

1. Is there a cost of reproduction, and how should it be measured?
2. Assuming there is such a cost, how often and how much should organisms optimally reproduce (often referred to as reproductive schedules)? This question may include age at first reproduction, frequency of reproduction, allocation of resources to each bout of reproduction and how these are affected by environment—overall favorability, fluctuations, density, and so on. In all of these investigations, we will use models that assume that genetic variability exists and that there are no constraints on reaching optima (known as phenotypic optimality models). This assumption is obviously not true, and there are more detailed models that incorporate details of genetic structure. The point of phenotypic models, especially at the very coarse scale intended here, is to predict general patterns among species and thus suggest what empiricists should be measuring to help explain patterns (Reznick 1985, 1992; Rose et al. 1987).

Cost of Reproduction

There has been a great deal of attention to the cost of reproduction in the past, for both plants (Ashman 1994) and animals (Roff 1992, Stearns 1992), although the problem has not been examined systematically in microbes, despite its obvious importance there. Reznick (1985) summarizes the literature into four general categories of evidence for a cost of reproduction:

1. phenotypic correlations
2. phenotypic with experimental manipulation of current reproduction and monitoring of subsequent survival and fecundity
3. genetic correlations
4. artificial selection experiments

Phenotypic approaches give an idea only of intrinsic physiological constraints. If there is no genetic basis to these constraints, it is not a topic for analysis using the theory of natural selection. However, if there is a strong physiological constraint, it seems a reasonable assumption that the constraints are pleiotropically related (i.e., the same gene complexes are responsible for increased reproduction and decreased allocation to maintenance). There are many studies showing a negative correlation between fecundity in year x with survival in year $x+1$, as would be expected by a tradeoff model. However, there are also many studies that do not show this pattern, and, indeed, some studies show the opposite. This variability underscores the problems inherent in trying to elucidate deeper meanings from simple correlations. Frequently, it seems to be the case, at least with sessile organisms, that an individual in a bad microsite has both low fecundity and low survivorship, while an individual in a good microsite has both high fecundity and high survivorship. Consequently, an empirical correlation shows a positive relationship between fecundity and survivorship.

In recent years, many studies have experimentally manipulated phenotypes, and in these too the results are variable (Ashman 1994). The underlying theory strongly suggests that there should be a substantial cost to reproduction, yet frequently in experimental situations that expectation is not met. Ashman argues that part of the problem is the difficulty in measuring investment in current reproduction; it is not actually fecundity that determines the cost but the total investment in reproduction (e.g., cost of flowers, nectar, etc.). In evaluating particular studies, it is important to take the following factors into account:

1. What is included in the studies—just offspring or all support, attractants, male function, etc.?
2. What current reproductive structures contribute to fecundity?
3. What currency is being used? Biomass is traditional, but nutrients may be more appropriate.
4. What is the initial investment (e.g., total carbon or nutrients initially put into structures) and what is actually lost (i.e., much is retrieved by resorption—e.g., Ashman [1994] shows values of 40% to 60% of nitrogen and phosphorus resorbed from floral structures)?
5. What time scale is used in the study?
6. What morphogenetic constraints might be involved?
7. What, if any, plasticity is relevant? Both phenotypic correlations and manipulative studies tend to assess phenotypic plasticity, which may or may not have a genetic component associated with it and therefore may or may not directly reflect evolutionary trends.

Studies of genetic correlations and artificial selection experiments have not been especially common, despite the fact that they are probably the best way of evaluating costs of reproduction in the context of deducing potential for evolutionary change. Perhaps the classic case is that of Rose and Charlesworth's (1981) natural selection experiments with *Drosophila melanogaster*, in which two lines were selected for. In line O, the flies were not allowed to successfully breed until later in life; in line B they were allowed to successfully breed only early in life. The expectation was that O flies would evolve to be late breeders and B flies, early breeders. O flies turned out not only to be later breeders than B flies; they also evolved the capacity to live longer (and therefore demonstrate a genetic correlation between length of life and time of breeding). The "cost" in this case of early reproduction was shortened life span in addition to the expected later breeding (see figure 2.1 for the data on egg production).

In addition to their importance in evolutionary theory, these questions may have important practical consequences. For example, Jack-

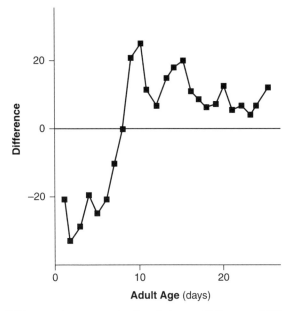

Figure 2.1. Difference in egg production between O lines and B lines from Rose and Charlesworth's (1981) experiment.

son (1980) has been promoting the idea that it would be useful for natural systems agriculture to breed grasses that are both perennial and high yielding (since natural prairies are based on perennial grasses, this is a clear goal resulting from the philosophy of natural systems agriculture). But the basic evolutionary theory of tradeoffs between reproduction and maintenance would suggest that this might be an impossible goal. The cost of maintaining rootstocks permanently would work against allocating more energy to the seeds (which is what we want to harvest). This constraint would explain why all domesticated grasses thus far are annuals. But recently Jackson and Dewald (1994) demonstrated that in the perennial grass *Tripsacum dactyloides* there appears to be a compartmentalization of energy allocation such that significant changes in energy allocation to reproduction can occur with no changes in allocation to maintenance activity (the compartmentalization appears to occur because of the autonomous development of reproductive tillers). This factor may suggest that the assumptions of a simple tradeoff model are too simplistic (Lande 1982).

Optimal Reproductive Schedules

Assuming that, despite the difficulty in measuring, there is a cost to reproduction, what is the optimal reproductive schedule? A plant or animal devotes some energy to reproductive activities and the rest to maintenance activities. How much energy should be allocated to reproduction versus maintenance during a given season? The most elementary formulation has no structure associated with it and involves the simple exponential map, namely,

$$N_{t+1} = \lambda N_t$$

where λ has been substituted for the R of chapter 1 (so as to conform with standard practice in the literature), and N is the density or biomass of the population. The parameter λ can be broken into two parts as follows,

$$\lambda = b + p$$

where b is per capita births and p is probability of survival. One can conceive of an organism as having to "decide" (short for natural selection acting on the problem) how much energy to devote to b, reproduction, versus other functions that will enhance p. This means that both b and p must be cast as functions of something else. Following the formulation of Schaffer and Gadgil (1975), we allow both b and p to be functions of energy devoted to reproduction. So we write,

$$\lambda = b(E) + p(E)$$

where E is the fraction of total energy devoted to reproduction. How might we expect b and p to vary with E? In general b increases and p decreases with E. (At the extreme, if $E = 100\%$, p has to be 0 since there is nothing left for maintenance because it has all been used up in reproduction; note that b and p have opposite behaviors with respect to E.) The expectation is that natural selection will act to maximize the value of λ: that is, the sum of b plus p. Using this model Schaffer (1974) discusses two general situations; either both functions are concave or both are convex. These two situations are shown in figure 2.2.

The result of this simple thought experiment is that we expect there

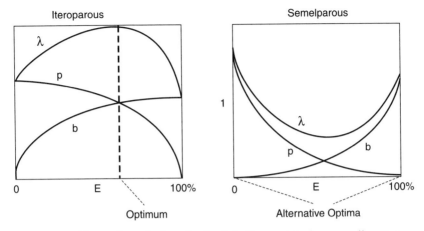

Figure 2.2. Alternative solutions for the Schaffer model of energy allocation.

to be two qualitatively distinct reproductive strategies in nature. In one strategy, the organism puts everything into maintenance or everything into reproduction (the semelparous strategy). In the other strategy, the organism puts some energy into reproduction and some into maintenance every year (the iteroparous strategy). These strategies, of course, correspond to what happpens in the real world. Here we have a model that suggests an underlying mechanism. If both curves are convex, there is a maximum at an intermediate E, which is the iteroparous solution. When both are concave the maximum is either at 0 or 100%, which is the semelparous solution.

Given these two dramatically different solutions that emerge from the underlying shape of the functions, it is useful to ask what ecological factors give rise to either concave or convex functions. Arguments usually focus on the fecundity curve because survival can be thought of as everything else. A convex fecundity function implies that there are diminishing returns for the energy devoted to reproduction. For example, if competition for pollinators decreased the proportion of flowers pollinated as more flowers were produced, a convex fecundity function would result. This means that while the total *number* of pollinated flowers continues to increase with increasing energy devoted to reproduction, the *probability* that any given flower will be pollinated declines. Or, to take an animal example, if birth weight becomes smaller with more individuals produced (as it must in many animals, such as mammals) such that spontaneous abortions occur with large litter expectancies, a convex fecundity would again result.

On the other hand, a concave function results from an accelerated accumulation of births with increasing energy devoted to reproduction. For instance, if the production of large numbers of flowers on a single plant attracted disproportionately more pollinators than a lower number of flowers, reproduction would increase nonlinearly with more energy devoted to reproduction (i.e., producing more flowers). In this case, not only are more flowers produced with more energy allocated to reproduction, but each flower has a higher probability of being pollinated. Another example might arise in a situation in which seed predators are important in the system. With low numbers of seeds produced not many seedlings are produced because almost all the seeds are eaten by seed predators. However, when seed production goes up, there is a tendency for the seed predators to become satiated, thus creating the critical nonlinearity (Silvertown and Dodd 1996).

As a general rule, we can say that some sort of facilitation is involved when there is a concave function while some sort of intraspecific competition is involved when the function is convex. An interesting correlation was found in the scarlet gilia (*Ipomopsis aggregata*) (Paige and Whitham 1987). At low elevations, this species is a classic semelparous species. But at higher elevations, it sometimes can be iteroparous. The presumed reason for this switch is the lower abundance of pollinators at higher elevations, which would lead to more competition for pollinators. That would tend to generate a convex curve. Paige and Whitham (1987) garnered further evidence for this mechanism by showing that more iteroparity was observed when they excluded pollinators from plants. However, they also found a greater tendency for iteroparity when they removed flower buds early in the season, a treatment that should have reduced the competition for pollinators.

If it is the case that we are dealing with an iteroparous situation, finding the optimal energy level to allocate for reproduction is simple. Given the basic equation for per capita population growth rate,

$$\lambda = b(E) + p(E)$$

we wish to find the maximum value of λ, with respect to E. Thus we compute,

$$\partial\lambda/\partial E = \partial b/\partial E + \partial p/\partial E$$

and set $\partial\lambda/\partial E = 0$ (in addition, the second derivative must be negative to ensure that it is a maximum rather than a minimum), to obtain,

$$\partial b/\partial E = -\partial p/\partial E$$

and we conclude that when the rate of change of birth rate with respect to reproductive energy is equal to the (negative) rate of change of survivorship with respect to reproductive energy, the per capita population growth rate (and individual fitness) is optimized.

The basic model can be modified in various ways to ask particular questions. For example, what might be the effect of variable environments on the optimal strategy? Suppose we have two environments that occur in a coarse-grained fashion (that is, a given year is either a good year or a bad year). The birth rate can be thought of as being multiplied by an environmental factor such that $b(1 + s)$ is the birth rate in a good year and $b(1 - s)$ is the birth rate in a bad year (where s is a measure of the quality of the environment). Then we have,

$$\lambda_g = b(1 + s) + p$$

$$\lambda_b = b(1 - s) + p$$

where the subscript g refers to good years and the subscript b refers to bad years. If we presume that the good years and bad years occur with the same frequency, we can express the overall population growth as,

$$\lambda^* = \lambda_g\lambda_b = [b(1 + s) + p][b(1 - s) + p] = b^2 + 2bp + p^2 - b^2s^2$$

(we multiply the rates because the variability in the environment occurs in a coarse-grained fashion). The derivative becomes,

$$\partial\lambda^*/\partial E = 2b(\partial b/\partial E) + 2p(\partial b/\partial E) + 2b(\partial p/\partial E) + 2p(\partial p/\partial E) - 2s^2(\partial b/\partial E)$$

Setting the derivative equal to zero and rearranging, we obtain,

$$[1 - s^2b/(b + p)][\partial b/\partial E] = -\partial p/\partial E$$

and since $[1 - s^2 b/(b+p)]$ must be a negative term for the equation to hold, we conclude that energy devoted to reproduction should decrease when variable environments affect birth rates. It pays to reproduce at a lower rate if doing so increases the chance of surviving to reproduce again, a phenomenon that has come to be known as bet hedging.

An important question arises when dealing with this sort of theory. What, precisely, should one measure in nature to test the theory? The theory says quite unambiguously that one should measure b and p as a function of E, the proportion of total energy devoted to reproduction. However, because one can never know all of what a plant or animal is doing physiologically, it is difficult to be certain that one is measuring all the components of reproduction. The best that can be done in field studies is to find surrogates for the components of the model. For example, Schaffer used the number of seeds per pod (i.e., the probability of being pollinated, an index of fecundity, which would most likely be correlated with λ) as a function of inflorescence size (a surrogate for energy spent in reproduction) for several iteroparous and semelparous species. The model predicts that there should be a relationship between these two variables for semelparous species but not for iteroparous species. Schaffer found that the relationship between seeds per pod and inflorescence size was indeed correlated significantly for semelparous but not for interoparous ones, precisely as predicted by the model.

A similar analysis can shed light on when a plant is predicted to be an annual versus a perennial (Schaffer and Gadgil 1975). For an annual plant we can write,

$$N_{t+1} = c(bN_t)$$

where b is the number of offspring per individual, as before, and thus bN_t is the total number of juveniles produced, and c is the juvenile survival probability. A perennial plant carries over some of the adults from year to year and thus can be represented as,

$$N_{t+1} = c(bN_t) + pN_t$$

where p is the proportion of adults surviving into the next time period. We thus see that the difference between the two strategies is simply setting the adult survival rate to zero (the definition of annual). So for the annual plant we can write,

$$\lambda_a = cb_a$$

and for the perennial

$$\lambda_p = cb_p + p$$

and using the previous logic that natural selection maximizes λ, we conclude that a plant will be a perennial (that is, the condition for $\lambda p > \lambda a$) when,

$$cb_p + p > cb_a$$

which can be rewritten as,

$$b_a > b_p + p/c$$

This forces us to conclude that fecundity for an annual strategy has to be higher than fecundity for a perennial strategy for the annual to be the favored strategy. Furthermore, because energy devoted to reproduction is usually lower for perennials (both empirically true and theoretically plausible since it takes more for maintenance; but see previous discussion), if p is approximately equal to c we have

$$b_a > b_p + 1$$

which says that all an annual has to do is produce one more seed than a perennial and it will have a higher fitness. We leave the reader with the question: Why then are there any perennials?

3

Projection Matrices:
Structured Models

Elementary Population Projection Matrices

Most populations are divided into different classes. Insects have eggs, larvae, pupae, and adults. Plants have seeds, seedlings, saplings, and adults. Models of populations in which the individuals are thus "structured" are referred to as structured population models, the most common form of which are projection matrices. There are two approaches to looking at the dynamics of populations, with either structured or unstructured models. In unstructured models (discussed in chapters 1 and 2), with or without density dependence, we assume that all individuals are equal. Unstructured models are most often used to make sense of changes in population sizes over time. Recent analyses of this sort of data have become extremely sophisticated (e.g., Hastings et al. 1993). Structured models, on the other hand, are often used to ask more detailed questions about populations, relating frequently to their management, their response to natural selection, or a variety of other specific issues.

Dividing individuals into classes can have dramatic effects, at least in the short term, on the overall dynamics of the population. A population of 10,000 eggs of some pest species has quite a different significance than a population of 10,000 adults, and a tree population with all seedlings is not usually called a forest, the name we would give to a tree population composed of mainly adult trees. Unfortunately, taking account of age or stage classes complicates matters considerably.

The most general structure in populations is age. All populations are composed of individuals that are young or old, and the variable used

---

Populations of trees contain dramatically different kinds of individuals, from seedlings to large adults. Population models must take this kind of structure into account to be realistic.

to classify the individuals is conveniently the same variable used to view the changes in the population, time. A 1-year-old child this year will be 2 years old next year and 11 years old 10 years from now.

The basic dynamics with two age classes, X and Y, can be visualized as follows:

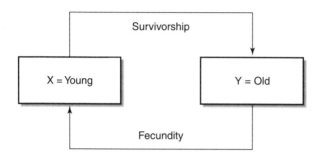

The simple projection map for a population (as introduced in chapter 1), can be expanded to include various age classes. So, rather than the simple,

$$N_{t+1} = aN_t$$

where N stands for the whole population, we have the two equations:

$$X_{t+1} = mY_t \tag{1a}$$

$$Y_{t+1} = pX_t \tag{1b}$$

where X is the population density of the youngest age class and Y is the population density of the oldest age class (and $N = X + Y$); on the assumption that we can divide the population into just two age classes, m is the number of young individuals produced by each old individual (the fecundity), and p is the probability that a young individual will survive to be an old individual (survivorship). Equations 1a and 1b can be conveniently represented as a matrix equation (readers who have forgotten their matrix algebra may consult Appendix A),

$$\begin{vmatrix} X_{t+1} \\ Y_{t+1} \end{vmatrix} = \begin{vmatrix} 0 & m \\ p & 0 \end{vmatrix} \begin{vmatrix} X_t \\ Y_t \end{vmatrix}$$

or in the more compact matrix form (following the customary procedure of printing vectors and matrices in boldface type),

$$\mathbf{N}_{t+1} = \mathbf{P}\mathbf{N}_t \tag{2}$$

where,

$$\mathbf{N}_t = \begin{vmatrix} X_t \\ Y_t \end{vmatrix}$$

and

$$\mathbf{P} = \begin{vmatrix} 0 & m \\ p & 0 \end{vmatrix}$$

Equation 2 is simply another way of writing equations 1a and 1b. The vector $\mathbf{N}$ is commonly referred to as the age distribution vector, and the matrix $\mathbf{P}$ is called either the projection matrix or the Leslie matrix, named after the person who is generally credited with the development of the method (Leslie 1945), although the procedure had actually been published three years earlier (Lewis 1942).

Of course, most populations structured into stages or ages will have more than just two classes, but this way of representing the population is easily generalizable. For example, letting X_0 represent the population density of the youngest individuals, X_1 that of the 1-year olds, X_2 that of the 2-year-olds, and so forth, up to X_n the population density of the n-year-olds (assuming n is the oldest the individuals in this population can get), equations 1a and 1b can be simply expanded, as follows,

$$X_0(t+1) = m_1 X_1(t) + m_2 X_2(t) + m_3 X_3(t) + \ldots + m_n X_n(t)$$

$$X_1(t+1) = p_{1,0} X_0(t)$$

$$X_2(t+1) = p_{2,1} X_1(t)$$

$$X_3(t+1) = p_{3,2} X_2(t)$$

.

.

.

$$X_n(t+1) = p_{n,n-1} X_{n-1}(t)$$

where the time variable is put in parentheses so as not to confuse it with the subscript that indicates the age category. This relatively large set of equations can be rewritten exactly as before, in the form of equation 2, where here the age distribution vector includes all of the Xs from X_0 to X_n, and the matrix $\mathbf{P}$ is

$$\mathbf{P} = \begin{vmatrix} 0 & m_1 & m_2 & m_3 & \cdots & m_n \\ p_{1,0} & 0 & 0 & 0 & \cdots & 0 \\ 0 & p_{2,1} & 0 & 0 & \cdots & 0 \\ 0 & 0 & p_{3,2} & 0 & \cdots & 0 \\ \cdot & & & & & \\ \cdot & & & & & \\ \cdot & & & & & \\ 0 & \cdots & & p_{n,n-1} & & 0 \end{vmatrix}$$

The first subdiagonal contains the survivorship probabilities, and the first row the fecundities. As in the previous case we can write,

$$\mathbf{N}_{t+1} = \mathbf{P}\mathbf{N}_t$$

which is identical to equation 2.

Note the similarity between equation 2 and the exponential map. They are identical except equation 2 deals with matrices and vectors. Indeed, the two equations are really two sides of the same coin, and the exponential map is simply a special case of equation 2 (the case with a single age category). However, an additional complication arises with the introduction of age categories into the model. Consider, for example, a population with three age categories, X_1, X_2, and X_3. Suppose the projection matrix is,

$$\mathbf{P} = \begin{vmatrix} 0 & 5 & 10 \\ 0.5 & 0 & 0 \\ 0 & 0.2 & 0 \end{vmatrix}$$

and the population begins with 10 individuals, all in category X_3 (for example, an island has just been invaded by 10 old individuals of the population). What will be the history of the population for the next few years? Applying equation 2, we obtain, for successive years,

$$\begin{vmatrix} 0 \\ 0 \\ 10 \end{vmatrix} \rightarrow \begin{vmatrix} 100 \\ 0 \\ 0 \end{vmatrix} \rightarrow \begin{vmatrix} 0 \\ 50 \\ 0 \end{vmatrix} \rightarrow \begin{vmatrix} 250 \\ 0 \\ 10 \end{vmatrix} \rightarrow \begin{vmatrix} 100 \\ 125 \\ 0 \end{vmatrix} \rightarrow \begin{vmatrix} 625 \\ 50 \\ 25 \end{vmatrix} \rightarrow \begin{vmatrix} 500 \\ 312 \\ 10 \end{vmatrix} \rightarrow \begin{vmatrix} 1662 \\ 250 \\ 62 \end{vmatrix}$$

and the total population for those 7 years will have been 10, 50, 260, 225, 700, 822, and 1975. The proportion of the total population in a particular age class varies enormously. For example, the oldest age class begins by representing 100% of the population (10 old individuals out of a total of 10 individuals), then 0%, then 0% again, then 4% (10/260), then 0%, then 4% again, then 1%, and finally 3%. If the projection is carried further, we find that this percentage stabilizes at around 2%. The same calculations could be done for the other age classes, making clear that even though there is great initial variation in the proportion of the population represented by each of the age classes, after some time the proportions stabilize. In this particular example, the young age class stabilizes at about 76%, the second category at about 22%. So if we imagine an alternative example of initiating a population of 10 individuals in which the distribution of individuals is

$$\begin{vmatrix} 7.6 \\ 2.2 \\ 0.2 \end{vmatrix}$$

(granted you cannot have fractional individuals, but this is just an example), the future populations look like this:

$$\begin{vmatrix} 7.6 \\ 2.2 \\ 0.2 \end{vmatrix} \rightarrow \begin{vmatrix} 13 \\ 4 \\ 0.4 \end{vmatrix} \rightarrow \begin{vmatrix} 23 \\ 6 \\ 0.76 \end{vmatrix} \rightarrow \begin{vmatrix} 40 \\ 12 \\ 1 \end{vmatrix} \rightarrow \begin{vmatrix} 72 \\ 20 \\ 2 \end{vmatrix} \rightarrow \begin{vmatrix} 124 \\ 36 \\ 4 \end{vmatrix} \rightarrow \begin{vmatrix} 219 \\ 62 \\ 7 \end{vmatrix} \rightarrow \begin{vmatrix} 381 \\ 109 \\ 12 \end{vmatrix}$$

We see that each age category is increasing, but if you calculate the percentage representation of each age category in the population as a whole, it remains perfectly constant or, loosely speaking, stable. This form of distribution of individuals in a population is referred to as a stable age distribution. Furthermore, because each age category is increasing by the same proportional amount, each of the separate equations must have the same constant rate of increase, which is to say,

$$X_i(t+1) = aX_i(t)$$

for all values of i, which means we can write,

$$\mathbf{N}_{t+1} = a\mathbf{N}$$

and furthermore, if N (not boldfaced) is equal to the total population,

$$N_{t+1} = aN_t$$

which is identical to the exponential map. In figure 3.1, the above examples are plotted over time. It is evident that the population beginning with a stable age distribution increases in a smooth exponential fashion, as normally expected, while the population beginning with 10 individuals in the oldest category has significant fluctuations before increasing in a smooth, typically exponential, fashion.

It is a well-known rule that, under normal circumstances, any population with constant mortality and natality will eventually reach a stable age distribution, at which point it will be growing according to the classic exponential equation. (There are technical exceptions to this rule, but they need not concern us presently. See Caswell 2001 for a more detailed discussion.) As discussed previously, most populations do not grow according to the exponential for long, and some form of density dependence usually sets in, or some controlling factor from above (a parasite or a predator, for example) limits the population. Introducing density dependence in an age- or stage-distributed model is a relatively complicated affair, as discussed later.

Any structured population at its stable age distribution increases in an exponential fashion. We can quite easily, with the aid of a computer, take such a population and simply project it into the future and calculate the rate at which it is growing. That rate will be essentially the same as we have seen before, the intrinsic rate of natural increase. But there is another way of calculating the rate of increase directly from the matrix. Mathematically speaking, the finite rate of increase is the same as the dominant eigenvalue of the projection matrix.

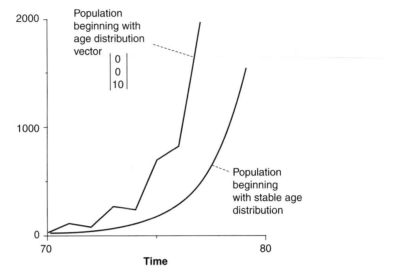

Figure 3.1. Population trajectories for different starting points in an age-distributed population.

As noted earlier, a population with n age categories will have a matrix that looks like

$$
\mathbf{P} = \begin{vmatrix}
0 & m_1 & m_2 & m_3 & \cdot & \cdot & \cdot & m_n \\
p_{1,0} & 0 & 0 & 0 & & \cdot & \cdot & \cdot & 0 \\
0 & p_{2,1} & 0 & 0 & & \cdot & \cdot & \cdot & 0 \\
0 & 0 & p_{3,2} & 0 & & \cdot & \cdot & \cdot & 0 \\
\cdot & & & & & & & \\
\cdot & & & & & & & \\
\cdot & & & & & & & \\
0 & \cdot & \cdot & \cdot & & p_{n,n-1} & & 0
\end{vmatrix}
$$

but the basic process of population projection remains,

$$
\mathbf{N}_{t+1} = \mathbf{P}\mathbf{N}_t \tag{3}
$$

We can also write,

$$
\mathbf{N}_{t+2} = \mathbf{P}\mathbf{N}_{t+1} \tag{4}
$$

Substituting equation 3 into equation 4 we get,

$$N_{t+2} = PPN_t = P^2N_t$$

and in general,

$$N_{t+n} = P^nN \tag{5}$$

Multiplying P by itself n times (assuming n is relatively large), we eventually approach the matrix

$$P^n = \begin{vmatrix} \lambda^n & 0 & 0 & 0 & \cdots & 0 \\ 0 & \lambda^n & 0 & 0 & \cdots & 0 \\ 0 & 0 & \lambda^n & 0 & \cdots & 0 \\ 0 & 0 & 0 & \lambda^n & \cdots & 0 \\ \cdot & & & & & \cdot \\ \cdot & & & & & \cdot \\ \cdot & & & & & \\ 0 & & \cdots & & & \lambda^n \end{vmatrix}$$

where λ is the general symbol usually used and is the dominant eigenvalue of the matrix P. Since P^n is a diagonal matrix with λ^n on the principal diagonal, we can substitute λ^n for P^n in equation 5, whence we obtain,

$$N_{t+n} = \lambda^n N_t$$

where, remarkably, we can substitute a simple scalar for the entire projection matrix (i.e., λ^n is not a matrix but an ordinary scalar number). If we now set the beginning of the projection as $n = 1$, we have, approximately,

$$N_{t+1} = \lambda N_t \tag{6}$$

which is a fundamental property of age-distributed populations. But, remember, equation 6 is true only for that special column vector that was the result of projecting the population repeatedly such that the t in equation 6 is relatively large. That is, as long as t is large, equation 6 says that each time projection results in each age category multiplied by a constant, and consequently the whole population is also increased by that amount. This means that λ must be identical to e^r, as first discussed in chapter 1.

We are now prepared to go one step further and derive one of the most important equations in matrix algebra, and the one that enables us to directly calculate the dominant eigenvalue. Recall equation 6,

$$\mathbf{N}_{t+1} = \lambda \mathbf{N}_t$$

By definition (remember the original definition of what a projection matrix is),

$$\mathbf{N}_{t+1} = \mathbf{P}\mathbf{N}_t$$

Therefore,

$$\mathbf{P}\mathbf{N}_t = \lambda \mathbf{N}_t$$

which can be written

$$\mathbf{P}\mathbf{N}_t - \lambda \mathbf{N}_t = 0$$

which, in turn, can be factored so as to give

$$(\mathbf{P} - \lambda \mathbf{I})\mathbf{N}_t = 0 \tag{7}$$

(**I** refers to the identity matrix—a square matrix with 1s on the principal diagonal and 0s everywhere else. Again, if your matrix algebra is rusty, refer to Appendix A.) Consider equation 7 for an annual plant population that has only two age categories (recall equation 1), adults (who produce m juveniles each time unit but do not survive more than a single year) and juveniles (who have a survival rate of g). Equation 7, for this example, gives us a pair of simultaneous linear equations,

$$\begin{vmatrix} -\lambda & m \\ g & -\lambda \end{vmatrix} \begin{vmatrix} N_0 \\ N_1 \end{vmatrix} = 0$$

$$-\lambda N_0 + mN_1 = 0 \qquad\qquad (8a)$$

$$gN_0 - \lambda N_1 = 0 \qquad\qquad (8b)$$

From equation 8a we write

$$N_0 = \frac{mN_1}{\lambda}$$

and, substituting into equation 8b we have,

$$\frac{gmN_1}{\lambda} - \lambda N_1 = 0$$

which simplifies to,

$$\frac{\lambda^2 N_1 - gmN_1}{\lambda} = \frac{(\lambda^2 - gm)\,N_1}{\lambda} = 0$$

which is true, if and only if (presuming $N_1 > 0$),

$$\lambda^2 - gm = 0$$

Note that the quantity on the left-hand side is, by definition, the determinant of

$$\begin{vmatrix} -\lambda & m \\ g & -\lambda \end{vmatrix}$$

which gives us the very basic equation

$$\det(\mathbf{P} - \lambda\mathbf{I}) = 0$$

This equation is referred to as the characteristic equation of the matrix $\mathbf{P}$ and is the equation used to find the value of λ (actually there are two values here, but for an n-dimensional system there will be n values). So, in the present example,

$$\mathbf{P} = \begin{vmatrix} 0 & m \\ g & 0 \end{vmatrix}$$

$$\mathbf{P} - \lambda \mathbf{I} = \begin{vmatrix} -\lambda & m \\ g & -\lambda \end{vmatrix}$$

$$\det(\mathbf{P} - \lambda \mathbf{I}) = \lambda^2 - gm$$

and $\lambda^2 - gm = 0$ has two roots $(gm)^{1/2}$ and $(-gm)^{1/2}$. The roots of the characteristic equation of any square matrix are called the eigenvalues of the matrix. For a population projection matrix, the largest of those roots is the rate of growth of the population after it has been projected a number of times with the same projection matrix.

Non-age Structure: Stage Projection Matrices

In many situations, the use of a Leslie-Lewis matrix is very inconvenient. Consider, for example, a population of long-lived trees, say oak trees that live as long as 200 years. We could define ages as quarter centuries (25 years) and have an eight by eight projection matrix. But a great deal of interest actually happens in the seed and seedling stage, which are lumped into the first 25 years all as if they were part of the same age category. A seed, a seedling, and a 25-year-old sapling can hardly be considered equivalent. On the other hand, if we define the ages as 1 year intervals (which would make sense for the seeds that may live for a maximum of a single year), we are stuck with a 200 by 200 projection matrix, an unwieldy object to say nothing of the prospect of having to estimate the survival and fecundity of individuals that differ from one another by a single year (e.g., the difference between a 225-year-old oak and a 226-year-old oak tree is not likely to be due mainly to the 1 year age difference). Furthermore, for the variables that are important, survivorship and fecundity, after the seed and seedling stage, the age probably matters little. A 100-year-old tree that has been growing in the shade and reaches only 10 m in height probably has survivorship and fecundity properties similar to those of a 30-year-old tree that has been growing in the sun and has reached that same 10 m. That is, for fecundity and survivorship, age is not the variable that really matters. Similar observations could be made about insects with discrete life cycle stages, long-lived sessile animals such as corals, and any other population in which characteristics other than age are the most important characteristics in determining vital rates.

In these sorts of situations, structured models do not retain the characteristics of the original Leslie-Lewis formulation: that is, categorizing individuals according to their ages. In some ways this loss is unfortunate, because the age-structured projection matrix has a very convenient form (all zeros except the first subdiagonal and the first row). But in many other ways it is the form of the projection matrix equation that matters, and the details of the structure of the projection matrix are irrelevant. Thus, much of the mathematical manipulation is just as relevant for projection matrices based on stages as it is for matrices based on ages. Lefkovitch (1965) pioneered the technique of using stage projection matrices.

For example, a particular species of tropical tree is sought for its lumber. It is, of course, impossible to do an experiment to decide what harvesting regime allows for a sustainable yield of timber (i.e., how much can be harvested while still retaining a viable population in perpetuity) because such an experiment would take hundreds of years to complete. The only option is to develop a model of the population. Because it is a highly structured population (seeds are different from seedlings are different from saplings, and so on), the most obvious model is one based on stage classes.

Suppose the population is conveniently divided into four stages: seed, seedling, sapling, and adult. The identification of the stage is obvious for the seeds, but what about the other three? There is no well-defined difference among seedling, sapling, and adult. We must provide an arbitrary measure that distinguishes these three categories. For example, seedlings are all plants less than a meter tall, saplings are all plants between 1 m tall and 10 cm in breast-height diameter, and adults are all others. Suppose we do a census of the population and discover that in 1 hectare (ha) there are 10,000 seeds, 5000 seedlings, 100 saplings, and 50 adults. Thus, if we construct a population vector similar to what we did in the case of an age distribution vector, we have

$$\begin{vmatrix} 10,000 \\ 5000 \\ 100 \\ 50 \end{vmatrix}$$

Now suppose that we are able to mark all individuals in the population (even all the seeds, although what we would undoubtedly do in practice is mark a subsample), and then we return next year to sample again. We find that 9 out of 10 seeds died and the other 10% germinated, thus making the probability of germination equal to 0.1. We find that of the 5000 seedlings marked, 1000 are still seedlings, 5 became saplings (grew to a height of more than a meter), and the rest died. Thus, the probability of staying a seedling is $1000/5000 = 0.2$, and the probability of growing to become a sapling is $5/5000 = 0.001$. Of the saplings marked, we find that 30 died (probability of mortality $= 30/100 = 0.3$), 2 became adults (grew to the point where their breast-height diameter was greater than 10 cm) (probability of growing to adulthood $= 2/100 = 0.2$), and the rest (68) remained as saplings (probability of remaining as saplings $= 68/100 = 0.68$). Five of the adults fell over (the probability of surviving as an adult is $45/50 = 0.9$). Finally, we made a count of all the seeds (all of which must be new because seeds last only a year) and discovered 10,000 (which means that on average each of the 10 adult trees produced 1000 seeds). With these data, we can write a series of four equations for the projection of each of the stage classes, letting $N_0 =$ number of seeds, $N_1 =$ number of seedlings, $N_2 =$ number of saplings, and $N_3 =$ number of adults. Namely,

$$N_0(t+1) = 1000 \, N_4(t) \tag{9a}$$

$$N_1(t+1) = 0.1 N_0(t) + 0.2 N_1(t) \tag{9b}$$

$$N_2(t+1) = 0.001 N_1(t) + 0.68 N_2(t) \tag{9c}$$

$$N_3(t+1) = 0.2 N_2(t) + 0.9 N_3(t) \tag{9d}$$

Solving these four equations, we find that the population of the second year is easily computed from the population of the first year. Furthermore, equations 9a, 9b, 9c, and 9d can be rewritten in matrix form as,

$$\begin{vmatrix} N_0(t+1) \\ N_1(t+1) \\ N_2(t+1) \\ N_3(t+1) \end{vmatrix} = \begin{vmatrix} 0 & 0 & 0 & 1000 \\ 0.1 & 0.2 & 0 & 0 \\ 0 & 0.001 & 0.68 & 0 \\ 0 & 0 & 0.2 & 0.9 \end{vmatrix} \begin{vmatrix} N_0(t) \\ N_1(t) \\ N_2(t) \\ N_3(t) \end{vmatrix}$$

or

$$\mathbf{N}_{t+1} = \mathbf{PN}_t$$

where, as before, boldface indicates matrices and vectors. Once again, the projection matrix $\mathbf{P}$ is a square matrix, but this time there are more entries than the simple subdiagonal and first row of the classic age projection matrix. If we repeatedly apply the projection matrix to the initial stage distribution vector, we obtain the series of vectors as follows:

$$\begin{vmatrix} 10000 \\ 5000 \\ 100 \\ 50 \end{vmatrix} \rightarrow \begin{vmatrix} 50000 \\ 2000 \\ 73 \\ 65 \end{vmatrix} \rightarrow \begin{vmatrix} 65000 \\ 5400 \\ 52 \\ 73 \end{vmatrix} \rightarrow \begin{vmatrix} 73100 \\ 7580 \\ 40 \\ 76 \end{vmatrix} \rightarrow \begin{vmatrix} 76118 \\ 8826 \\ 35 \\ 77 \end{vmatrix} \rightarrow \begin{vmatrix} 76609 \\ 9377 \\ 33 \\ 76 \end{vmatrix} \rightarrow$$

The population appears to be growing, and, especially if you follow the adult trees only (which an ecosystem manager or forester is likely to do) you could come to the conclusion that the population is indeed in a healthy state, perhaps stabilizing at about 76 adult trees per hectare. But if we apply the matrix repeatedly, after 200 years we get the pattern shown in figure 3.2A. Clearly, this was not a viable population. It may take 200 years, but with the vital statistics in matrix $\mathbf{P}$, we can expect that the population will be effectively extinct after a couple of centuries.

This example provides an interesting exercise in the use of such models in ecosystem management (assuming that the numbers above are actual numbers of individuals in a real population that can be observed). A forester who sees the pattern after 5 years of observation of this population may conclude not only that this is a healthy population but also that it could be beneficially harvested. Since it changed from

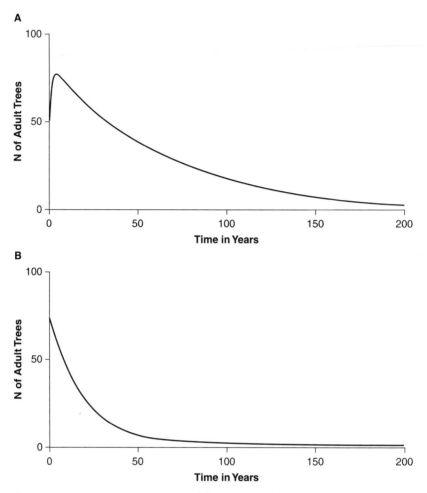

Figure 3.2. Long-term projections of the tree population discussed in the text. (A) After 5 years, the population appears to be growing but then begins a rapid decline to become very small after 200 years. (B) After the harvesting regime has been changed, on the basis of observations of the initial growth period (as shown in part A), the population declines even more rapidly, becoming virtually extinct after only 100 years.

only 50 trees per hectare to more than 75 trees per hectare in only 5 years, one might conclude that harvesting some of the adult trees certainly could not have a big effect on the population. So the forester may conclude that a 5% harvest of the standing stock (adult trees) would certainly not hurt the population (i.e., with 50 trees per hectare only 2.5

trees per year would be harvested, and with 75 trees per hectare only 3.75 trees per year would be harvested). It is easy to see how one might think of such a minimal intervention as not significant for the overall health of the population. But changing the element $p_{3,2}$ in the **P** matrix from 0.9 to 0.85 and making the repeated projections, we obtain:

$$
\begin{vmatrix} 76609 \\ 9377 \\ 33 \\ 76 \end{vmatrix} \rightarrow \begin{vmatrix} 76000 \\ 9536 \\ 32 \\ 71 \end{vmatrix} \rightarrow \begin{vmatrix} 71200 \\ 9507 \\ 31 \\ 67 \end{vmatrix} \rightarrow \begin{vmatrix} 66883 \\ 9021 \\ 31 \\ 63 \end{vmatrix} \rightarrow \begin{vmatrix} 63085 \\ 8493 \\ 30 \\ 60 \end{vmatrix} \rightarrow \begin{vmatrix} 59763 \\ 8007 \\ 29 \\ 57 \end{vmatrix} \rightarrow \begin{vmatrix} 56778 \\ 7577 \\ 28 \\ 54 \end{vmatrix}
$$

After just 6 years the adult population is almost back to where it was when the original observations were made (about 50 individuals). Furthermore, applying the **P** matrix (with the 0.9 probability changed to 0.85) we observe the pattern shown in figure 3.2B. The population is virtually extinct after only about 100 years. The harvesting intervention has had a large effect indeed and one that would not have been easy to foresee without the projection model.

The stage projection matrix is obviously a far more general model than that of the original age projection model. Yet the basic rules of matrix manipulation apply to stage projection matrices as well. Specifically, if we have,

$$\mathbf{N}_{t+1} = \mathbf{PN}_t \tag{10}$$

we can also write

$$\mathbf{N}_t = \mathbf{P}^t \mathbf{N}_0$$

Any square matrix multiplied by itself a large number of times results in a matrix with its dominant eigenvalue raised to the power of t on its principal diagonal and zeros everywhere else, approximately, which means we can write,

$$\mathbf{N}_t = \lambda^t \mathbf{N}_0$$

which also means that

$$N_{t+1} = \lambda N_t \qquad (11)$$

which will be true after the population has settled down to a stable stage distribution (the same as a stable age distribution except we are here referring to stages rather than ages). Since the left-hand sides of equations 10 and 11 are identical, we can equate the right-hand sides to obtain,

$$PN_t = \lambda N_t$$

which can also be written

$$PN_t - \lambda N_t = 0$$

or

$$(P - \lambda I)N_t = 0$$

which, as explained in a previous section, is only true when

$$\det(P - \lambda I) = 0 \qquad (12)$$

As explained before, this is the characteristic equation that is used to find the eigenvalues of the projection matrix.

It is important to realize that when populations are modeled as stages (these models are frequently referred to as i-stage models), there is really no restriction on where the transition probabilities must appear in the matrix. That is, in the strictly age-structured case all transition probabilities are survivorship probabilities, and all appear precisely in the first subdiagonal of the matrix. But in the case of other structures defining the stage (size, morphological distinctness, life cycle stage, and so on), depending on the actual biology involved, probabilities can occur in almost any part of the matrix. However, there is a certain pattern. If the stages are ordered in some biologically interesting way (e.g., small to big trees, early to late instars), the subdiagonals refer to the probability of advancing stages, the supradiagonals refer to the probability of regressing, and the principal diagonal contains the probabilities of remaining in the stage (see figure 3.3). It may seem unusual for individuals to "regress" stages, but such behavior can be

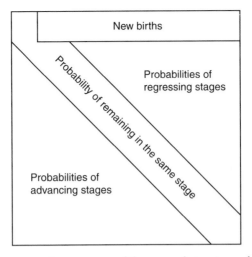

Figure 3.3. Diagrammatic summary of the general structure of a stage projection matrix.

common, depending on how stages are defined. For example, a tree population is usually modeled with size as the categorization variable. It is quite common, especially with seedlings and young saplings, for that size to sometimes decrease: deer browse oak seedlings; large leaves fall and cover tropical tree seedlings; killing their upper branches but allowing lower branches to survive and form new vegetation. Many such examples could be cited.

Eigenvectors, Reproductive Value, Sensitivity, and Elasticity

Recall the basic equation, now repeated at least twice in this chapter,

$$\mathbf{N}_{t+1} = \lambda \mathbf{N}_t$$

As discussed previously, when a population follows this equation we refer to its age distribution as the stable age distribution, meaning that the proportion of individuals in each age category does not change as the projection matrix continues to be applied. The related equation

$$\mathbf{PN}_t = \lambda \mathbf{N}_t \tag{13}$$

basically says the same thing. When equation 13 is true, the column vector **N** is referred to as an eigenvector (recall that λ is the eigenvalue), and we will write it as,

$$\mathbf{Pu} = \lambda\mathbf{u}$$

using **u** as a special symbol for the special column vector that satisfies equation 13. The vector **u** is thus the stable stage (or age) distribution vector. It is also the case that if we take the transpose of **P** (exchange the rows and columns of the matrix) and symbolize the transpose as **P***, we can write,

$$\mathbf{P}^*\mathbf{v} = \lambda\mathbf{v} \qquad (14)$$

where the symbol **v** is used as a special symbol for the vector that satisfies equation 14 (which itself is true when equation 13 is satisfied, which is to say when the population has reached its stable stage distribution). Exactly what is the meaning of this vector?

According the formal terminology of matrix algebra, when a vector satisfies equation 13, the parameter λ is the eigenvalue. The vector **u** is referred to as the right eigenvector and the vector **v** as the left eigenvector (the left eigenvector because **P***v** can also be written as **v***P**, where the asterisk refers to the transpose of the matrix—**v*** is a row matrix while **v** is a column matrix—and thus **v*** is multiplied on the left while **u** is multiplied on the right by the matrix **P**). Taking a simple example, we write out equation 13 as,

$$\begin{vmatrix} p_{1,1} & p_{1,2} & p_{1,3} \\ p_{2,1} & p_{2,2} & p_{2,3} \\ p_{3,1} & p_{3,2} & p_{3,3} \end{vmatrix} \begin{vmatrix} u_1 \\ u_2 \\ u_3 \end{vmatrix} = \begin{vmatrix} \lambda u_1 \\ \lambda u_2 \\ \lambda u_3 \end{vmatrix} \qquad (15)$$

and equation 14 would be,

$$\begin{vmatrix} p_{1,1} & p_{2,1} & p_{3,1} \\ p_{1,2} & p_{2,2} & p_{3,2} \\ p_{1,3} & p_{2,3} & p_{3,3} \end{vmatrix} \begin{vmatrix} v_1 \\ v_2 \\ v_3 \end{vmatrix} = \begin{vmatrix} \lambda v_1 \\ \lambda v_2 \\ \lambda v_3 \end{vmatrix} \qquad (16)$$

The difference between equations 15 and 16 is that the square matrix of 16 is the transpose of the square matrix of 15. We already know that the right eigenvector contains the elements of the stable age distribution. What might be the meaning of the left eigenvector, and might it have some sort of biological significance?

To answer this question, write out one of the linear equations of the matrix equation 15, say the second equation. We have,

$$p_{2,1}u_1 + p_{2,2}u_2 + p_{2,3}u_3 = \lambda u_2$$

which can also be written as,

$$u_2 = (p_{2,1}u_1 + p_{2,3}u_3)/(\lambda - p_{2,2}) \tag{17}$$

Note that on the left-hand side is the single element referring to the value of the second stage in the stable stage distribution, while on the right-hand side are all the probabilities that contribute to that stage (the first subscript on all the probabilities is 2). Now do the same multiplication for equation 16, obtaining,

$$p_{1,2}v_1 + p_{2,2}v_2 + p_{3,2}v_3 = \lambda v_2$$

which can also be written as,

$$v_2 = (p_{1,2}v_1 + p_{3,2}v_3)/(\lambda - p_{2,2}) \tag{18}$$

Here we have on the left-hand side the single element referring to the value of the second stage in the left eigenvector, while on the right-hand side are all the probabilities to which that stage contributes. In other words, v_2 represents the total contribution of the second stage category to the other categories (the second subscript on all the probabilities is 2). For this reason, the elements of the left eigenvector are called the reproductive value (not an especially good choice of term but one we seem to be stuck with from extensive use in the literature). As is evident from the difference between equations 17 and 18, the elements of the right eigenvector refer to the contribution of all other stages to that stage (the stable stage distribution), whereas the elements of the left eigenvector refer to the contribution of that stage to all other stages (the reproductive value).

Once a projection matrix has been estimated from real data, the basic information of interest to the population ecologist is the rate of population increase (the dominant eigenvalue, λ), the stable age distribution (the right eigenvector), and the distribution of reproductive values (the left eigenvector). But, although these figures may be of interest in and of themselves, the way they might change is often of interest also. An ecosystem manager who wishes to set size limits on a sport fishery or extraction limits on a nontimber forest product in a rain forest may want to know how the dominant eigenvalue changes in response to a management decision. The goal is to understand the effect of a management decision on the rate of population growth. Mathematically, we wish to know the value of

$$s_{ij} = \partial\lambda/\partial p_{ij}$$

where s_{ij} is referred to as the sensitivity of the matrix with respect to p_{ij}. The meaning of sensitivity is clear. How much will the rate of population growth change with a change in p_{ij}? The value of s_{ij} is given by the simple formula (see Caswell 1978 or 2001 for a derivation)

$$s_{ij} = v_i u_j \qquad (19)$$

where v_i and u_j are the appropriate elements of the left and right eigenvectors (the reproductive value of the ith stage times the stable stage element of the jth stage). As elegant a formula as 19 is, its utility is limited because of scaling problems. For example, transition probabilities are always between 0 and 1, whereas natality is frequently very large (thousands for some plants, millions for corals). So the change in λ from a "unit" change in p_{ij} can be very misleading. For this reason, it has been convenient to develop a "proportional" representation of the same idea, a measure that would give a fractional contribution to λ of a change in p_{ij}. Caswell and colleagues (1984) proposed the use of a concept borrowed from microeconomics, the elasticity of a parameter. The elasticity is given as the rate of change in the log of λ with respect to the log of an element of **P**. Specifically,

$$e_{ij} = \partial(\ln\lambda)/\partial(\ln p_{ij}) \qquad (20)$$

which [since $d\ln x = (1/x)dx$] is

$$e_{ij} = \frac{p_{ij}}{\lambda} \frac{\partial \lambda}{\partial p_{ij}} = \frac{p_{ij}}{\lambda} v_i u_j$$

(21)

which is the formula that can be used for calculating the elasticities (find the dominant eigenvalue and the two eigenvectors and do the appropriate multiplication). It is not at all obvious from the definition of elasticity (equation 20), but the sum of all elasticities in the whole matrix is equal to 1.0 (de Kroon et al. 1986). This fact means that we can interpret e_{ij} as the proportional sensitivity of λ to changes in p_{ij} (because all the es sum to 1.0 they are interpretable as proportions). However, it is frequently useful to look at both sensitivity and elasticity when interpreting a population. Sensitivity is especially useful when entries in the matrix are zero. With a zero entry, the elasticity is automatically 0 (see equation 21). This means that it is not possible to assess the importance of elements that are zero, even though a small change in one of them might result in a large change in λ. This problem is due to the original logarithmic definition of elasticity, equation 20.

Applications of Population Projection Matrices

The main problem with applying structured models in nature is estimating the values of the parameters. Unfortunately, it is very difficult to treat this subject in a general fashion because each organism has special features that need to be taken into account in estimating transition probabilities and fecundities. However, there are a couple of generalizations to be noted.

One basic feature of a structured model is the transition probability, which is most easily, or ideally, obtained by marking and following individuals over time. For example, if there are $N_i(t)$ individuals in stage i at time t, and $N_{i+1}(t+1)$ individuals in stage $i+1$ at time $t+1$, we *cannot* conclude that the transition probability from i to $i+1$ is the simple ratio $N_{i+1}(t+1)/N_i(t)$, unless the stage is strictly age (i.e., a Leslie-Lewis model). This would seem to be a self-evident fact. However, if we have $N_i(t)$ individuals marked (all in stage i at time t) so as to recognize those particular individuals in the future, and we find that of those $N_i(t)$ marked individuals, $N_{i+1}(t+1)$ of them show up at the next time interval in the $i+1$th stage, it seems that we may conclude that $P_{i,i+1}$ (the transition probability from stage i to stage $i+1$) is equal to the simple ratio $N_{i+1} (t+1)/N_i(t)$. Such a simple calculation is in fact not necessarily correct for some technical reasons that need not concern us at this point (see Vandermeer 1975). In fact, the calculation is dependent on the assumption that the population within the stage category i to $i+1$ is distributed according to a stable stage distribution. This assumption is most likely to be met if the definition of stage is as narrow as possible. However, making stage definitions narrow limits the amount of data available to make the estimate. This basic contradiction is inherent in any study attempting to instantiate a structured population model.

Estimation of transition probabilities and fecundities results in what is frequently referred to as a dynamic life table (or a cohort life table). An alternative approach, the static life table, uses a snapshot at one point in time of how many individuals are in each stage or age category and uses those numbers to estimate survivorship and reproduction.

The Dall's Mountain Sheep: A Static Life Table

Consider, for example, the classic case of the Dall's mountain sheep (*Ovis dalli*), originally analyzed by Murie (1944; see Deevey 1947 for a summary) and reproduced here as table 3.1. The number of individuals in each age category is estimated by the age of 608 carcasses found in Mount McKinley National Park (now Denali National Park). If we assume that survival and number of births are constant from year to year, we can calculate survival rates from these data. This assumption can be true only if the total number of births in the population is the same; that is, if λ is near 1 and survival rates are constant from year to year. Such assumptions are rarely met in natural populations, but frequently they are made so as to get at least some information about a critical population.

Such a life table is constructed by summing up all the death records and supposing that this number is equal to the number in the population at the beginning of a particular projection (this is the reason that the sum of numbers in column 2 is 608—the number surviving at the beginning of the first age interval). Then each category's death record is subtracted to produce the estimate of the number surviving at the beginning of the next age interval. Finally, each number in column 3 is divided by the original number in the cohort to produce the number surviving as a fraction of the original cohort. This is the classic way of

TABLE 3.1
Static Life Table Based on 608 Carcasses of Dall's Mountain Sheep Found by
Murie in Mount McKinley National Park in 1944

Age Interval in Years	Number Dying during Age Interval	Number Surviving at Beginning of Age Interval	Number Surviving as a Fraction of Original Cohort
0 to 1	121	608	1
1 to 2	7	487	0.801
2 to 3	8	480	0.789
3 to 4	7	472	0.776
4 to 5	18	465	0.764
5 to 6	28	447	0.734
6 to 7	29	419	0.688
7 to 8	42	390	0.640
8 to 9	80	348	0.574
9 to 10	114	268	0.439
10 to 11	95	154	0.256
11 to 12	55	59	0.096
12 to 13	2	4	0.006
13 to 14	2	2	0.003
14 to 15	0	0	0

elaborating a static life table and can be used to estimate survivorship
probabilities from stage to stage.

Such static life tables are of limited use because their construction is
dependent on the assumption that the population is at a stationary age
or stage distribution, which is to say the eigenvalue is 1.0 and the population is at a stable age or stage distribution. Dynamic, or cohort, life tables are generally more interesting because one can ask whether the
population is growing or declining, whether the stable age distribution
has been reached, what the elasticities are, and a host of other questions.

Palo de Mayo: A Dynamic Life Table

Consider the growth of a population of *Vochysia ferruginea*, a lowland
tropical rainforest tree. Boucher and Mallona (1997) reported on the
growth of a population of this species after a hurricane. The storm dev-
astated the entire forest of southeastern Nicaragua in 1988, so it might
be reasonable to expect that for the next 5 years the population would

experience its "maximum" possible growth rate, which is to say it would be free of density-dependent effects. Classifying the population into five stages (seedling, small sapling, large sapling, small adult, and large adult), marking individuals, and following the same individuals over a 5-year period, Boucher and Mallona constructed the following projection matrix:

	Seedling	Small sapling	Large sapling	Small adult	Large adult
Seedling	0.209	0	0	35.6	70.1
Small sapling	0.010	0.653	0.020	0	0
Large sapling	0	0.170	0.407	0	0
Small adult	0	0	0.570	0.731	0
Large adult	0	0	0	0.266	0.997

From this projection matrix they calculated that $\lambda = 1.156$. This result has important practical consequences. Immediately after the hurricane, forest managers were concerned that the devastating storm could have caused a local extinction of the species. Because it is an important species for the timber industry, such a speculation was cause for concern. However, with the calculation of $\lambda = 1.156$, it became clear not only that the population was healthy but also that, projecting into the future, the population will dominate the forest by 2014. Obviously, such a projection is unrealistic since our assumption about density independence cannot last forever. Nevertheless, the simple calculation of the dominant eigenvalue put to rest the concerns of local forest managers.

From this projection matrix Boucher and Mallona calculated the following sensitivity matrix,

	Seedling	Small sapling	Large sapling	Small adult	Large adult
Seedling	0.09	0	0	0	0
Small sapling	8.47	0.17	0.04	0.05	0.09
Large sapling	25.11	0.51	0.12	0.16	0.26
Small adult	32.72	0.07	0.15	0.20	0.34
Large adult	40.12	0.82	0.19	0.25	0.42

and the following elasticity matrix.

	Seedling	Small sapling	Large sapling	Small adult	Large adult
Seedling	0.017	0	0	0.017	0.057
Small sapling	0.075	0.098	0.0007	0	0
Large sapling	0	0.075	0.041	0	0
Small adult	0	0	0.075	0.128	0
Large adult	0	0	0	0.057	0.358

From the elasticity matrix we see the overwhelming importance of the survivorship of the adult trees ($12.8 + 35.8 = 48.6\%$ of the value of λ is due to these two stages alone). This information is certainly of concern to foresters as they begin making decisions about harvesting this tree for timber. Note that examining the elasticities does not enable us to ask some potentially interesting questions. For example, it is perfectly possible that because of the nature of this forest the probability of going from a small adult to a large sapling or from a large adult to a small adult may become important in the future. (Falling debris becomes more important a factor as the forest matures, meaning that, frequently, small adults will suffer severe crown damage, effectively turning them into large saplings.) That is, individuals may very well regress stages in the future, something not reflected in the current matrix. The sensitivities can give us some information about the potential impact of this force. Sensitivity for the small adult to large sapling stage is 0.16 and that for the large adult to small adult 0.34, meaning a combined sensitivity (summing the two) of 0.5, which is larger than the sensitivity of the large adult survivorship (which, according to the elasticity analysis was the most important of all).

The American Beech: Testing Hypotheses with Dynamic Life Tables

The American beech (*Fagus grandifolia*) is a classic shade-tolerant climax species that occurs throughout northeastern North America. Despite the fact that its primary niche appears to be a climax forest, it persists as a dominant in hurricane-disturbed forests of the Atlantic and Gulf coasts. How can the species exist in this damage-prone area, where one would normally expect to find only pioneer-type vegeta-

TABLE 3.2
Population Trends Predicted by Four Hypotheses Proposed to Explain the
Long-Term Persistence of *Fagus grandifolia* (from Batista et al. 1998).

Hypothesis	Open Canopy	Closed Canopy	Long-Term Alternation
Recovery	Negative	Positive	Stable
Resistance	Stable	Stable	Stable
Release	Positive	Negative	Stable
Complementation	Negative	Negative	Stable

Note: The open-canopy phase starts with a hurricane and includes the time when a large fraction of the forest is in multiple-treefall gaps. The closed-canopy phase occurs after the canopy has been restored from the most recent hurricane damage and a comparatively large fraction of the forest is under canopy or in single-treefall gaps.

tion, when it is so obviously a primary forest species? There appear to be four reasonable hypotheses to explain its persistence: First, it may actually decline in numbers after disturbance but recover after the closed-canopy phase. Second, it may be generally resistant, stable in both phases and unaffected by the disturbance. Third, it may be a species that actually requires disturbance for establishment, as many species do. Normally, isolated light gaps in the forest are thought to provide the disturbance habitat, but this effect could be especially important after a damaging storm. Fourth, the species may regenerate only in the open phase and reproduce only in the closed phase, thus needing both phases of the damage cycle. The four hypotheses generate different predictions about how the population will respond in open versus closed canopy situations, as summarized in table 3.2.

These four hypotheses were tested with long-term demographic data gathered between 1978 and 1992, interrupted by hurricane Kate in 1984. It is clear that individual-level data are strongly influenced by the hurricane. Large-tree mortality went up drastically during and just after the hurricane, for example. But how was the population growth affected? Clearly, the best way to answer the question is to wait a few hundred years to see what happens, but some of us are too impatient and want an answer sooner. For us, the best approach is to develop a projection matrix and project the population numerically into the future. From their long-term census data, Batista and coworkers (1998) calculated two projection matrices, as shown in table 3.3. Casual examination of the demographic figures (i.e., the transition probabilities) reveals clear differences from before to after the hurricane, but it is not immediately obvious what the population-level consequences might be. Batista and coworkers then calculated the eigenvalue for the matrix as a whole and the elasticities for the three general components of survival, growth, and fecundity. These values are shown in table 3.4. The first and most important thing to note is that the eigenvalues are almost identical (statistically neither is different from 1.00). This similarity strongly suggests that the resistance hypothesis is most likely cor-

TABLE 3.3

Transition Matrices Estimated for the *Fagus grandifolia* Population in Woodyard Hammock in the Closed- (1978–1984) and Open- (1984–1992) Canopy Phases

Diameter at breast height (cm)	1978–1984								
	2–4	4–6	6–11	11–16	16–28	28–40	40–52	52–64	64–
2–4	0.70	0	0	0	0.04	0.15	0.23	0.25	0.25
4–6	0.16	0.69	0	0	0	0	0	0	0
6–11	0	0.23	0.84	0	0	0	0	0	0
11–16	0	0	0.11	0.81	0	0	0	0	0
16–28	0	0	0	0.15	0.88	0	0	0	0
28–40	0	0	0	0	0.08	0.84	0	0	0
40–52	0	0	0	0	0	0.11	0.80	0	0
52–64	0	0	0	0	0	0	0.14	0.76	0
64–	0	0	0	0	0	0	0	0.17	0.98
	1984–1992								
2–4	0.47	0	0	0	0.06	0.23	0.38	0.42	0.42
4–6	0.43	0.33	0	0	0	0	0	0	0
6–11	0	0.54	0.65	0	0	0	0	0	0
11–16	0	0	0.26	0.61	0	0	0	0	0
16–28	0	0	0	0.33	0.83	0	0	0	0
28–40	0	0	0	0	0.12	0.83	0	0	0
40–52	0	0	0	0	0	0.13	0.80	0	0
52–64	0	0	0	0	0	0	0.12	0.74	0
64–	0	0	0	0	0	0	0	0.11	0.49

TABLE 3.4
Asymptotic Population Growth Rates Projected with the Transition Matrices
Estimated for the Closed- (1978–1984) and Open- (1984–1992) Canopy Phases

		Elasticity Analysis (fraction of contribution to λ)		
Period	λ	Survival	Growth	Fecundity
1978–1984	0.98	0.83	0.14	0.02
1984–1992	1.01	0.71	0.25	0.04

rect (see table 3.2). But it is also of interest to look at the elasticities. Survival has a greater elasticity than growth, which has a greater elasticity than fecundity (a pattern that is quite general in plants), but the relative sizes of the elasticities are different in the two populations. Growth and fecundity contribute relatively more to the eigenvalue of the posthurricane forest than to that of the prehurricane forest, while survival contributes more in the prehurricane forest. Thus, although the species appears to be resistant (in the sense of table 3.2), the demographic characteristics are such that the persistence of the population in the two different circumstances is at least partially different; *Fagus grandifolia* has two different life styles at the population level.

Density Dependence in Structured Populations

Thus far all of the material in this chapter has been developed under the assumption that vital statistics are not dependent on the density of the population. We have tacitly assumed that the forces acting on an oak seedling are the same in an open field as in the understory of an oak forest, that a lone coral hydroid on a cement block in the middle of the ocean has a survivorship similar to that of one in the middle of the Great Barrier Reef, that a bacterial cell isolated in the middle of a nutrient agar petri dish is equivalent to a bacterial cell in the middle of a petri dish that is spilling over with bacteria. All are ridiculous assumptions. Yet they are reasonable under certain circumstances. For example, we might just want to know the current rate of population growth (or decline) for conservation purposes. The density-independent assumption is perfectly appropriate, although it would be folly to attempt a long-term projection. Or, in some cases, populations do behave as if they were density independent, at least for short periods of time, and modeling them over those short periods of time is reasonable.

However, if we wish to understand the behavior of a population over a long period of time, we must modify the density-independent assumption. In the same way we had to modify the exponential equation to take into account the obvious forces of density dependence, so must we modify it for structured populations. Unfortunately, things become very complicated very quickly.

Density Dependence in a Simple Age-Structured Model

Consider a population in two distinct phases, say, larva (X) and adult (Y). If we presume that both larva and adult live exactly one time period (one time period is necessary for larval development and metamorphosis, and all adults die after one time period), the projection matrix will be as discussed above:

$$\begin{vmatrix} X_{t+1} \\ Y_{t+1} \end{vmatrix} = \begin{vmatrix} 0 & m \\ p & 0 \end{vmatrix} \begin{vmatrix} X_t \\ Y_t \end{vmatrix} \tag{22}$$

As explained earlier, a projection matrix with constant coefficients always represents an exponentially growing population, yet most populations do not grow exponentially. To incorporate density-dependent effects, just as we developed the logistic equation in the previous chapter, we expand the model (equation 22) to allow for nonlinearities (density-dependent effects) in both the fecundity factor (m) and the survival probability (p). Assume the survivorship factor is

$$p = r_1(1 - X_t)$$

which is to say that the larvae undergo a logistic form of survivorship. Maximum survivorship is at the limit as X_t approaches zero, and survivorship decreases as a function of the population density of the larvae (r_1 must be less than or equal to 1.0, since p is a probability). Assume that the fecundity factor is

$$m = r_2(K - Y_t)$$

where we presume an upper limit, more or less a carrying capacity, on the population of potential offspring (K), and a maximum rate of offspring production of r_2K. Equation 22 thus becomes,

$$\begin{vmatrix} X_{t+1} \\ Y_{t+1} \end{vmatrix} = \begin{vmatrix} 0 & r_2(K-Y_t) \\ r_1(1-X_t) & 0 \end{vmatrix} \begin{vmatrix} X_t \\ Y_t \end{vmatrix}$$

Multiplying this matrix equation, we obtain the two-dimensional map (two dimensional because we have two state variables, X and Y):

$$X_{t+1} = r_2 Y_t(K - Y_t) \qquad (23a)$$

$$Y_{t+1} = r_1 X_t(1 - X_t) \qquad (23b)$$

where, for technical reasons beyond the scope of this chapter (i.e., the model blows up if we don't), we presume,

$$(1/4) < K < (2[r_2]^{-1/2})$$

This map is conveniently represented as a one-dimensional map through the process of composition, namely, beginning with

$$X_{t+1} = g(Y_t) = r_2 Y_t(K - Y_t)$$

$$Y_{t+1} = f(X_t) = r_1 X_t(1 - X_t)$$

we substitute $f(X_t)$ into $g(Y_t)$. That is,

$$X_{t+2} = g(Y_{t+1}) = g[f(X_t)]$$

This is the mathematical process of composition, and with the specific logistic functional forms considered here, we have

$$X_{t+2} = g[f(X_t)] = r_2 f(X_t)[(K - f(X_t)]$$

which is,

$$X_{t+2} = r_1 r_2 X_t(1 - X_t)[K - r_1 X_t(1 - X_t)] \qquad (24)$$

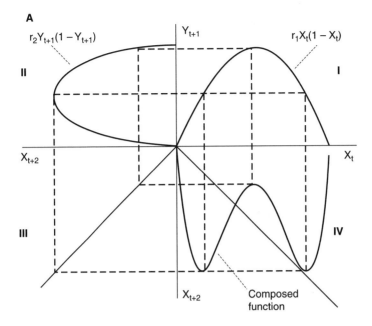

A

$r_2Y_{t+1}(1 - Y_{t+1})$ Y_{t+1} $r_1X_t(1 - X_t)$

II   I

X_{t+2} X_t

III   IV

X_{t+2} Composed
function

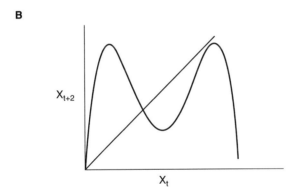

B

X_{t+2}

X_t

Figure 3.4. The process of composition.

which is a quartic function (i.e., involves a variable raised to the 4th power), the qualitative dynamics of which are reasonably well known (Vandermeer 1997). With both of these equations nonlinear, it is possible for the composed function itself to be monotonic, or to have one or two "humps." In figure 3.4 this process of composition and its result are illustrated. In the first quadrant (labeled I in figure 3.4A), the first

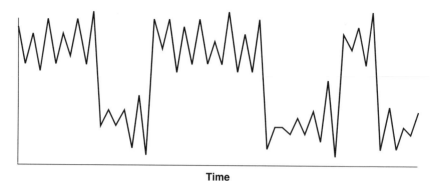

Time

Figure 3.5. Exemplary population trajectory based on the model of figure 3.4. Note the unpredictable nature of the pattern.

function (equation 23b) is plotted. In the second quadrant (labeled II in figure 3.4A) the second function (equation 23a with $K=1$) is plotted on its side to make the axis Y_{t+1} fit on both the ordinate of quadrant I and the abscissa of quadrant II, making it easy to visualize the projection from X_t to Y_{t+1} to X_{t+2}. The third quadrant (labeled III in figure 3.4A) simply maps X_{t+2} into itself, so as to generate the values of X_{t+2} as a function of X_t, the function shown, upside down, in quadrant IV. Turning the function of quadrant IV over (figure 3.4B), we have a complete graph of the composed function (equation 24 with $K=1$).

Projecting a population with this map leads to a great variety of interesting patterns (Vandermeer 1997), one of which is shown in figure 3.5. We return to this example for a more complete explanation in chapter 4.

Thus, we see how complex things can become when we add nonlinearities to the projection matrix. And this was only with two age categories. If a similar procedure were applied to populations with more age categories, with simple quadratic functions modeling the density-dependent effects at each age category, the resulting function may have many more humps. Indeed the general rule is the number of potential humps is 2^{n-1}, where n is the number of age categories. The possibilities for extremely complicated behavior of the population through time are enormous, as discussed in the next chapter.

Density Dependence in Size-Distributed Populations

The above discussion was made relatively simple by the assumption that we were dealing with an age-distributed population. We now turn

PROJECTION MATRICES **85**

to size-distributed populations. In this section, we look at some empirical examples that emphasize the importance of size in population dynamics, and in the section that follows we deal with the rather more
complicated problem of modeling density-dependence in size- (or
more generally stage-) distributed populations.

In chapter 2 we analyzed a variety of approaches to intraspecific
competition (density dependence), ranging from the logistic equation
to yield–density relationships to the self-thinning law. Here, we expand those approaches in the context of a stage-distributed population, focusing on size as a stage characteristic.

First, size per se is clearly important for the fate of individuals,
independent of competitive interactions. As far as we know, the largest
individual organism living in the world today is not a blue whale as
popularly believed but an individual of the fungus *Armillaria gallica*,
estimated to weigh more than 100 tons (also estimated to be 1500 years
old). And this individual began as a single cell. The difference between
a single cell and a 100-ton individual can hardly be ignored. In the case
of plants, small individuals are relatively more likely to die them are
large ones because of their inability to secure resources or their failure
to withstand physical forces. Similarly, smaller individuals usually reproduce less. In the case of animals, a change in size may result in a
change in habitat, effective availability of food, food type eaten, or risk
of predation, among other factors. In fungi, the total biomass of hy-
phae is most likely related to reproductive output. Similar statements
could be made for the Phaeophyta (brown algae) and Rhodophyta (red
algae). Thus, size is an important issue for population dynamics of at
least these five kingdoms, yet it is not really covered in classical theory.

Second, although it is almost always the case that organisms get
larger as they become older, variation in size at a given age is also important both in how it is generated and in its consequences. At a most
general level we can write

$$S(t+1) = rS(t) \tag{25}$$

where $S(t)$ is size at time t and r is relative growth rate of the individual. Since this is an exponential form, it cannot be true for an extended
period of time, but for initial growth of individuals it is frequently a
convenient shorthand. A variety of ecological factors influence the relative growth rate, but the time of emergence (or birth) and the size of
the egg (seed) both influence the size attained at a particular time. All

else being equal, initial size should increase size at any subsequent time. This is a general rule that is frequently assumed, especially in evolutionary studies. However, it is worth noting that this rule assumes no confounding effect between relative growth rate and initial size, which does not seem to be the case in plants (e.g., Gross 1984).

Size structure has been an important theme both in plant and animal populations. Frequently, the focus is more on variation in size because even if total population biomass is the same, the distribution of individual sizes may be different (a billion redwood seedlings may have the same biomass as a population with a single adult). In a study of the limpet *Patella cochlear*, Branch (1975) determined that as the population grows, the average size of individuals decreases, such that high-density populations have a large number of small individuals and only a few large ones, whereas low-density populations have a large number of large individuals and only a few small ones (see figure 3.6). Furthermore, there has been a convergence on the general conclusion that populations with greater variation in size are actually more stable in total numbers over time. Why might this be the case?

In simple dynamic models, such as equation 25, we can look at dynamics as output–input graphs. Equation 25 is linear (i.e., the log of size increases as a straight line as a function of time—the exponential model), but more often than not the relationship over large density or size ranges is nonlinear. At higher population densities or at higher sizes, relatively fewer offspring are produced. In general, as discussed previously, we can write,

$$N(t+1) = f[N(t)]$$

where the function f can be very nonlinear with remarkably complicated results as will be discussed in chapter 4. For now, suffice it to say that a function that curves downward at high densities means that very high density populations actually produce fewer propagules (eggs, seeds, etc.) than intermediate densities. Depending on how extreme this difference is, it can lead to population cycles.

What density has to do with variation in size is perhaps not obvious. However, one likely reason for the decline in output at very high input is a lack of size variation combined with nonlinear effects of size on reproduction. If all individuals are affected equally by low resources at high density, at a very high density all individuals might be too small to reproduce

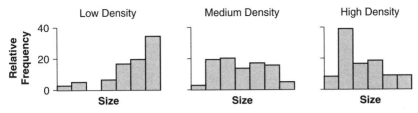

Figure 3.6. Intraspecific competition and growth in populations of the limpet *Patella cochlear*. High-density populations have many small and a few large individuals; low-density populations have many large and a few small individuals (Branch 1975).

effectively. On the other hand, if there are strong size hierarchies, as the hierarchy gets stronger at high density, there will always be a few large individuals that reproduce adequately and keep total output from declining. Thus, it is easy to see that a strong size hierarchy in annual plants (and perhaps other organisms as well) tends to stabilize a population over the long run. Given this generalization, it is clearly important to understand what exactly controls the degree of variation in size among individuals.

We begin with the problem of quantifying size variation. Most plant populations not only exhibit variation in size among individuals but also exhibit a particular form of that variation: skewed with many small and a few large individuals. Many other organisms show this pattern also: e.g., corals, wasps, brown algae, red algae). Older literature used the variance and skewness coefficient to characterize size variability, but newer literature uses the Gini coefficient (Weiner 1986), borrowed from economics, where it is a measure of inequality of distribution. The Gini coefficient is calculated as,

$$G = \frac{\sum_{i=1}^{n} \sum_{j=1}^{n} \left| x_i - x_j \right|}{2\bar{x}n\,(n-1)}$$

Where x_i is the size of the ith individual in the population, x bar is the mean of the sizes, and n is the population density. The meaning of the Gini coefficient can be grasped most easily from a graph of distribution of wealth (its original utility in economics). In figure 3.7, we illustrate a graph of the distribution of wealth in the Netherlands at the present time compared with the distribution of wealth in Guatemala. In Guatemala, almost 80% of the country's wealth is owned by only 10% of the people, a staggering level of inequality. In the Netherlands, only 20% of the wealth is owned by 10% of the people, still not equal but more so than in Guatemala. The area between the actual curve of wealth distribution and the theoretical egalitarian situation (a straight line) is equal to the Gini coefficient. The Gini coefficient takes into account both variability and skewness.

Early literature in plant population ecology often assumed that the existence of size hierarchies was evidence of competition among individuals. However, even populations of plants not in competition with one another will develop a size hierarchy, simply because of the exponential nature of early plant growth in biomass (Turner and Rabinowitz 1983), as described by equation 25. If individuals vary even slightly in initial size (or emergence time or relative growth rate), any individual that gets a slight head start will grow at a larger absolute rate, even if at the same per unit size rate, and the difference in size among individuals will continue to increase, during the exponential growth phase.

Although competition is not necessarily the cause of size hierarchies, it can influence the rate at which hierarchies develop. This influence takes on three distinct forms. First, larger plants may take up a disproportionate share of resources. That is, relative growth rate increases with size, so that, as density increases, the initially bigger plants get a disproportionate advantage. They thus get much larger, while even slightly smaller plants do not get as much of an advantage. In this way, individuals become more and more unequal in their absolute size. This process results in the Gini coefficient's increasing with density at a given time or over time at a given density (Weiner and Thomas 1986).

Second, if each individual receives a share of resources proportional to its size, bigger plants get more resources. Thus, all plants get a proportional share of resources. Relative growth rate is constant with size, and thus, as density increases every plant gets proportionately less resource and grows more slowly. Because the coefficient of variation (or Gini coefficient) increases with size (i.e., the original distribution spreads out), if all individuals grow more slowly, there is less variability at higher densities. This is the opposite of what is expected with disproportionate resource utilization.

These two forms of resource uptake (proportionate to biomass and disproportionate to biomass) are parallel to what the traditional literature refers to as size-asymmetric (also called one-sided competition)

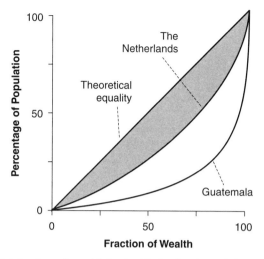

Figure 3.7. Distribution of wealth in the Netherlands and Guatemala. Shaded area is the Gini coefficient for the Netherlands.

and size-symmetric (also called two-sided competition), usually related to whether the competition is for light or for nutrients. When competition is for light, it is apparent that larger individuals get more than their proportionate share of the resource (they intercept all the light and cast shade on their smaller brethren, who get very little). When competition is for belowground resources, a bigger individual will have a proportionately bigger root system and will more likely consume an amount of nutrient in proportion to its overall size. In a classic experiment, Weiner (1986) used a simple design to separate out effects of one-sided and two-sided competition. His results are illustrated in figure 3.8.

The third way in which competition influences the rate at which hierarchies develop has to do with mortality. If the smallest plants are more likely to die (almost always true), this relation between size and mortality will tend to truncate the distribution and thus generate less inequality among surviving plants. So, as competition gets more intense and mortality begins happening (self-thinning), as described in chapter 1, competition results in the reverse expectation from size-symmetric competition. As a general rule, competition generates mortality more in the case of competition for light. This is a simple physiological process in that when a plant is shaded by a competitor such that

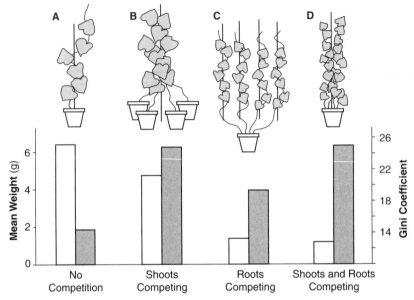

Figure 3.8. An experiment on vines (*Ipomoea tricolor*) in which root and shoot competition were separated. Mean weight (open bars) was significantly different ($p < 0.001$) for all comparisons between treatments except (C) and (D). The Gini coefficient in weight (shaded bars) was significantly different for the comparison of treatments A and B ($p < 0.05$) and of treatments A and D ($p < 0.01$) (Weiner 1986, 1990).

it is living below its compensation point, it must die. Competition for nutrients generally results in smaller individuals, but rarely leads to death.

The above can be summarized as a sequence of mechanisms in time. Assume we start with an even-aged monoculture (e.g., a tree plantation or a field of crop plants). Initially, all plants are small, and any competition between plants is probably competition for nutrients and thus a proportional type. Consequently, higher initial density should lead to less inequality. As the canopy begins closing, competition becomes disproportionate and the generation of a size hierarchy is accelerated. Finally, with further canopy closure some plants experience light regimes that are below their compensation points and thus die, now reversing the tendency to form a size hierarchy. Obviously, this pattern will depend on particular habitats. For example, a particularly low-nutrient situation may never proceed to the canopy closure stage (e.g., deserts).

To this point, we have, according to tradition in plant ecology, confounded two different concepts of size-symmetry generation—resource uptake and growth rate—implicitly using size-symmetry of growth rate to indicate size-symmetry of uptake. Recently, Schwinning and Weiner (1998) distinguished among various categories on the basis of resource uptake. Their categories are reproduced in table 3.5.

The terms *size-asymmetric* and *size-symmetric* competition are normally applied to plant populations. A very similar idea has been applied to animal populations, but here it is called *scramble competition* versus *contest competition*. In scramble competition, individuals in a population obtain and utilize resources in proportion to their ability to do so (frequently based on their size). In contest competition, some individuals (frequently the bigger ones) win a disproportionate amount of resource as they "contest" with others in the population. Thus, scramble competition is similar to (perhaps ultimately the same as) symmetric competition, and contest competition is similar to asymmetric competition.

TABLE 3.5
Definitions of Competitive Mechanisms Generating Size Hierarchies
(Schwinning and Weiner 1998)

Term	Definition
Complete symmetry	All plants receive the same amount of resource, irrespective of their sizes
Partial size symmetry	Uptake of contested resources increases with size, but less than proportionally
Perfect size symmetry	Uptake of contested resources is proportional to size (equal uptake per unit size)
Partial size asymmetry	Uptake of contested resources increases with size, and larger plants receive a disproportioate share
Complete size asymmetry	The larger plant gets all the contested resources

Density Dependence in a Stage-Structured Model

Recently, a team of mathematicians and biologists collaborated on a series of studies attempting to predict the population behavior of the flour beetle *Tribolium castaneum* (Costantino et al. 1997). Their results have been quite spectacular thus far and warrant introduction here. A fuller presentation of their results will be undertaken in chapter 4, after some more advanced analytical techniques have been introduced. We make the introduction here simply as an example of the way in which density dependence can be incorporated into a stage-structured model (and the complications that arise when that incorporation is attempted). The analysis of the model necessarily must wait for some additional analytical tools, which are developed in chapter 4.

Tribolium populations can be divided into three stages: larva, pupa, and adult. The effect of adults on larvae is complicated because adults both "produce" larvae (the egg stage is not included in this model) and consume larvae. Cannibalism occurs, with adults eating both pupae and eggs and larvae eating eggs. Thus we expect a density-dependent effect through this cannibalism, where the production of larvae will be decreased by both more larvae and more adults (since both adults and larvae eat eggs). From the basic biology, it is also reasonable to assume that (if the time step is appropriately chosen) larvae do not remain as larvae from one time unit to the next, nor do pupae, but adults do. Thus, the number of larvae at some time point is a function of the number of larvae and adults at the previous time, the number of pupae is a function of the number of larvae at the previous time, and the number of adults is a function of the number of pupae (some of which will turn into adults, others of which will be eaten by the adults) and the number of adults (the ones that survive). The overall projection model, then, looks like,

$$
\begin{vmatrix} L_{t+1} \\ P_{t+1} \\ A_{t+1} \end{vmatrix} = \begin{vmatrix} 0 & 0 & f_1(L_t, A_t) \\ p_{lp} & 0 & 0 \\ 0 & f_2(A_t) & p_{aa} \end{vmatrix} \begin{vmatrix} L_t \\ P_t \\ A_t \end{vmatrix}
$$

where L is the number of larvae, P is the number of pupae, and A is the number of adults. The functions f_1 and f_2 stipulate the nonlinear effect of cannibalism on the population. Costantino and coworkers stipulated the functions as:

$$f_1 = \frac{b}{e^{c_1 L_t + c_2 A_t}}$$

and

$$f_2 = \frac{1}{e^{c_3 A_t}}$$

where the meanings of the constants c_1, c_2, and c_3 are not of particular importance here. Suffice it to say that this simple model defies simple analysis. Its behavior is remarkably complex, yet with some modern tools of analysis, to be introduced in the next chapter, some sense can be made of the model and, remarkably, the flour beetles seem to behave very much as the model predicts.

Basic Matrix Manipulations

Matrix Multiplication

A matrix is a table of numbers, and the mathematical manipulation of matrices derives from analysis of systems of linear equations. Thus, for example, if we have the system,

$$Y_1 = a_1 X_1 + b_1 X_2 + c_1 X_3 \tag{A1a}$$

$$Y_2 = a_2 X_1 + b_2 X_2 + c_2 X_3 \tag{A1b}$$

$$Y_3 = a_3 X_1 + b_3 X_2 + c_3 X_3 \tag{A1c}$$

simply as a matter of convenience we can group the X_is to the right and more easily visualize the structure of the system as follows,

$$
\begin{vmatrix} Y_1 \\ Y_2 \\ Y_3 \end{vmatrix}
=
\begin{vmatrix} a_1 & b_1 & c_1 \\ a_2 & b_2 & c_2 \\ a_3 & b_3 & c_3 \end{vmatrix}
\begin{vmatrix} X_1 \\ X_2 \\ X_3 \end{vmatrix}
$$

Here we have three matrices: first, a matrix with a single column (the Ys); second, a square matrix with three columns (sometimes referred to as the detached coefficient matrix); and third, a matrix with

94

a single column (the Xs). Sometimes the various matrices are simply referred to with a single letter, but it is customary when speaking of matrices to put them in boldfaced type, so the above equation could be

$$\mathbf{Y} = \mathbf{AX} \qquad \text{(A2)}$$

Equation A2 is identical to the system A1, except it obviously is more compact. This way of writing a set of linear equations becomes especially convenient when dealing with large systems. A system of 500 equations could be just as easily writtten as equation A2.

The way of interpreting A2 is the same as that for any other equation; the matrix $\mathbf{Y}$ (when there is only a single column, we sometimes refer to the matrix as a column vector) is equal to the column vector $\mathbf{X}$ multiplied by the matrix $\mathbf{A}$. But multiplying matrices is not as simple as multiplying scalars (nonmatrix variables—that is, regular numbers). We basically need to perform an operation such that we obtain the system A1 from the equation A2. The basic rule is, multiply each element in the first row of the square matrix by the corresponding element in the column vector and sum the results (i.e., $a_1X_1 + b_1X_2 + c_1X_3$). That gives us the first element in the $\mathbf{Y}$ vector (i.e., $Y_1 = a_1X_1 + b_1X_2 + c_1X_3$). Next, multiply each element in the second row of the square matrix by the corresponding element in the column vector and sum the results (i.e., $a_2X_1 + b_2X_2 + c_2X_3$). That gives us the second element in the $\mathbf{Y}$ vector (i.e., $Y_2 = a_2X_1 + b_2X_2 + c_2X_3$). Follow the same procedure to get the third element in the $\mathbf{Y}$ vector. Such is the relatively simple process of multiplying a column vector by a square matrix.

Frequently, it is necessary to multiply square matrices by square matrices. The process is really nothing more than multiplying the square matrix by successive column vectors. Take the simple example of a pair of two-by-two square matrices:

$$\begin{vmatrix} a_1 & b_1 \\ a_2 & b_2 \end{vmatrix} \begin{vmatrix} c_1 & d_1 \\ c_2 & d_2 \end{vmatrix}$$

Begin the process of multiplication first by simply considering the first column of the second matrix (that is the column with the cs) being multiplied by the square matrix to its left. Following the procedure outlined above we have $a_1c_1 + b_1c_2$ as the first element in the first column of the resultant matrix and $a_2c_1 + b_2c_2$ as the second element in the first column of the resultant matrix, or,

$$\begin{vmatrix} (a_1c_1 + b_1c_2) & \text{———} \\ (a_2c_1 + b_2c_2) & \text{———} \end{vmatrix}$$

where the horizontal lines indicate the calculations that are still to be made. Now repeat the process for the second column in the second matrix (the one with the ds), such that we get $a_1d_1 + b_1d_2$ for the first element in the second column of the resultant matrix and $a_2d_1 + b_2d_2$ for the second element in the second column of the resultant matrix, so that the final result is,

$$\begin{vmatrix} (a_1c_1 + b_1c_2) & (a_1d_1 + b_1d_2) \\ (a_2c_1 + b_2c_2) & (a_2d_1 + b_2d_2) \end{vmatrix}$$

A moment's reflection should convince the reader that premultiplication is not equivalent to postmultiplication (that is, in multiplying the two matrices **A** and **B** together you must specify whether you are multiplying **AB** or **BA**; the results will not generally be the same). Furthermore, it is possible to multiply matrices together only if the number of columns of the first matrix is equal to the number of rows of the second.

Matrix Addition and Subtraction

Adding and subtracting matrices is a far simpler affair than multiplication. Just add or subtract the equivalent elements in each matrix. Note that you can add or subtract only matrices that have the same number of rows and columns. The following example should make the process crystal clear. Subtract matrix **B** from matrix **A**.

$$\mathbf{A} = \begin{vmatrix} a_1 & b_1 \\ a_2 & b_2 \end{vmatrix}$$

$$\mathbf{B} = \begin{vmatrix} c_1 & d_1 \\ c_2 & d_2 \end{vmatrix}$$

$$\mathbf{A} - \mathbf{B} = \begin{vmatrix} (a_1 - c_1) & (b_1 - d_1) \\ (a_2 - c_2) & (b_2 - d_2) \end{vmatrix}$$

The Identity Matrix

Recall from basic arithmetic the identity element. That is, we need an element such that when we multiply it by any other element, the result is that same element again. That is,

$$aI = a$$

where I is the identity element. In elementary arithmetic we all recognize $I = 1$. Any number multiplied by 1 gives that element back again. A similar requirement exists in matrix algebra. That is, we need an element such that $\mathbf{AI} = \mathbf{A}$ (where $\mathbf{A}$ and $\mathbf{I}$ are matrices, as indicated by their bolded status). Recalling the basic rules of matrix multiplication, it does not take much thought to convince yourself that a matrix with 1s on the principal diagonal and 0s everywhere else will act as the identity element for any square matrix (with the same number of rows as the identity matrix). That is, consider the three-by-three matrix

$$\mathbf{A} = \begin{vmatrix} a_1 & b_1 & c_1 \\ a_2 & b_2 & c_2 \\ a_3 & b_3 & c_3 \end{vmatrix}$$

If we then perform the multiplication of **AI**, we obtain,

$$\mathbf{AI} = \begin{vmatrix} a_1 & b_1 & c_1 \\ a_2 & b_2 & c_2 \\ a_3 & b_3 & c_3 \end{vmatrix} \begin{vmatrix} 1 & 0 & 0 \\ 0 & 1 & 0 \\ 0 & 0 & 1 \end{vmatrix} = \begin{vmatrix} a_1 & b_1 & c_1 \\ a_2 & b_2 & c_2 \\ a_3 & b_3 & c_3 \end{vmatrix}$$

The Determinant of a Matrix

The determinant is an important concept the significance of which is difficult to convey in any sort of intuitive fashion. The concept derives from solving systems of linear equations. Consider, for example the following set of equations:

$$k_1 = a_1 X_1 + b_1 X_2 \tag{A3a}$$

$$k_2 = a_2 X_1 + b_2 X_2 \tag{A3b}$$

Solve the first equation for X_1:

$$X_1 = (k_1 / a_1) - (b_1 / a_1) X_2$$

Now substitute the value of X_1 (that is, the right-hand side of the above equation) into A3b, obtaining

$$k_2 = a_2 [(k_1 / a_1) - (b_1 / a_1) X_2] + b_2 X_2$$

and rearrange so as to put X_2 on the left-hand side:

$$X_2 = \frac{a_1 k_2 - a_2 k_1}{a_1 b_2 - a_2 b_1} \tag{A4}$$

Look closely at the denominator. If we arrange equations A3a and A3b as a matrix equation we get

$$\begin{vmatrix} k_1 \\ k_2 \end{vmatrix} = \begin{vmatrix} a_1 & b_1 \\ a_2 & b_2 \end{vmatrix} \begin{vmatrix} X_1 \\ X_2 \end{vmatrix}$$

If we now multiply the elements of the principal diagonal together (i.e., a_1b_2) and subtract from that the product of the two off-diagonal elements (a_2b_1), we have precisely the denominator of equation A4. This value is referred to as the determinant of the matrix and is symbolized as det **A**, for the determinant of the matrix **A**. So we have

$$\det \mathbf{A} = a_1b_2 - a_2b_1$$

where the matrix **A** is

$$\begin{vmatrix} a_1 & b_1 \\ a_2 & b_2 \end{vmatrix}$$

Now look at the numerator of A4. Now we form the matrix

$$\begin{vmatrix} a_1 & k_1 \\ a_2 & k_2 \end{vmatrix}$$

which is to say, we replace the second column of the detached coefficient matrix with the column of constants, k_1 and k_2. Recall the basic definition of the determinant—multiply the elements of the principal diagonal and subtract from the result the product of the off-diagonal elements. Thus, we obtain

$$\det \begin{vmatrix} a_1 & k_1 \\ b_1 & k_2 \end{vmatrix} = a_1k_2 - a_2k$$

which you will recognize as the numerator of equation A4. So we could generalize and say that if we form the matrix $\mathbf{A}_i$, where the ith column (in this case $i=$ either 1 or 2) of the detached coefficient matrix has been replaced with the column of constants (i.e., the elements of the column vector on the left-hand side of the equation), we have a general equation:

$$X_i = \det \mathbf{A}_i / \det \mathbf{A}$$

This equation is known as Cramer's rule (he was a Frenchman, so it is pronounced kra-MAYS rule). The amazing thing about it is that it applies to any system of linear equations. If you just solve an arbitrary system for one of the variables (say the fourth variable) you come up with a ratio that is the ratio of two determinants, the numerator being from the matrix that had its fourth column substituted with the column of constants and the denominator being the determinant of the detached coefficient matrix itself. Of course, calculating the determinant for a matrix larger than two by two is more complicated but is not really required of the reader of this book.

4

A Closer Look at the "Dynamics" in Population Dynamics

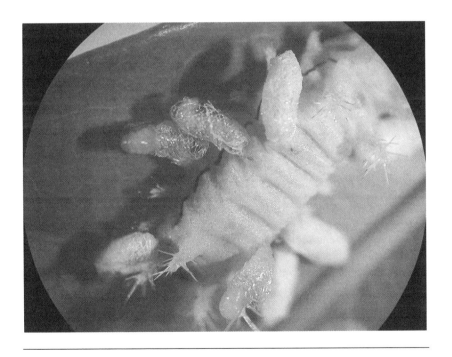

Central to the analysis of population dynamics are concepts that come most naturally to anyone trained in the classical sciences. In elementary physics classes, for example, a physical system is most frequently looked at from the point of view of stability and equilibrium. When engineers design systems, from aerospace to industrial management, one of the first questions asked is, Under what conditions will the system be at equilibrium, and will it be stable?" Naturally, ecologists began by asking such questions of ecosystems. As a consequence, concepts such as balance (equilibrium) and stability have become central to modern ecology, especially in population and community ecology. Often, such concepts have also become normative; they are viewed as goals to strive for in the design of sustainable resource management systems. Some ecosystems are reportedly unstable, having lost the inherent equilibria of natural ecosystems (Altieri 1987, Soule et al. 1990), and the job of good husbandry is to try to promote management tools that will restore balance and stability and therefore sustainability to the system (Altieri 1987, Levins and Vandermeer 1990). Such is a widely held position.

In population and community ecology the concepts of equilibrium and stability have been debated and clarified over the past 20 years, and their meaning has been operationalized to a considerable extent. This operationalization forced a rethinking of some central concepts. One example is the idea that the complexity of a system is related to its stability. As a system becomes larger, it develops more interconnec-

---

Parasitic wasps spin cocoons as they emerge from their moth larvae hosts. When the moth population is large, the wasp population can surge dramatically, causing a precipitous decline in the moths. These forces generate surprising complexity in the moth populations.

tions and, much as they do in a spider web, those interconnections make it resilient to outside perturbations and thus stable: the more interconnections a system has (read, the larger and more complex the system), the more stable it should be. However, the notion that a large highly connected system would be more stable than a small system with low connectivity was challenged by May (1974), who proved that, all else being equal, the larger and more complex the system, the more likely it is to be unstable, precisely the opposite of what most ecologists intuitively felt. Rather than being like spider webs, according to May, ecosystems are like houses of cards, deriving their structure from the myriad connections among parts but becoming more fragile the larger they grow.

The original intuition of ecologists was that large ecosystems, with their biodiversity and complex interconnections, are more stable and more in "balance" (at equilibrium) than simplified systems that had purposefully been designed to eliminate much of that complexity. It seems like an obvious idea. How can it be that careful analytical thought suggests otherwise? One of the problems, perhaps the principal problem, is that these early conceptualizations were based on classical notions of dynamics (stability, equilibrium, and balance) from the physical sciences. That basis coupled with the semiromantic notion of the balance of nature, a mainstay of nature lovers and environmental activists alike, and produced a concatenation of the popular with the scientific that created what seemed to be an unassailable principle. But there was a false analogy built up between the naturalists' notion of balance and stability and the classical engineers' notion of those same words. With more modern interpretations of the underlying dynamic structure of this and other concepts, largely derived from the new science of nonlinear dynamic systems, a new classification of dynamic behaviors may correspond better with the old naturalists' or traditional farmers' original intuition, as explained in the following sections.

Intuitive Ideas of Equilibrium and Stability

The intuitive notions of balance and stability have their parallels in classical analytical thought, balance as equilibrium and stability as either unstable or stable. Consider the graph in figure 4.1. The variable x could be any interesting variable, but for our purposes it is best to

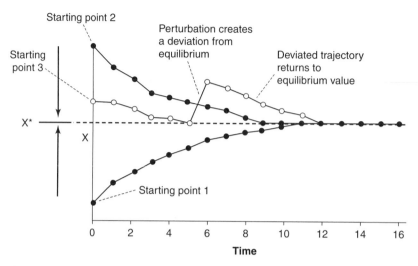

Figure 4.1. Illustration of the dynamic behavior associated with a point attractor.

think of it as population density. Plotting density over time, beginning at various starting points, we see that no matter where the trajectory begins, it always ends up at the value x^*. Furthermore, once it attains the value of x^* it never deviates. The value x^* is thus an equilibrium point (the system is "in balance" once it reaches that point). That it is in balance is only one feature of x^* that is important. The behavior of the variable when it is not exactly at that equilibrium point is also of great importance. Although it is true that when at equilibrium (the variable x exactly equals x^*) the system remains there in perpetuity, it is also true that the slightest deviation from that value (say, to the point labeled "deviation from equilibrium" in figure 4.1) results in a return to that same equilibrium. Because any such deviation results in a return to the same equilibrium point, the point is a "stable" equilibrium point.

In contrast, consider the situation presented in figure 4.2. Again, there is an equilibrium point (x^*), and again if the system is initiated at exactly that point, it will remain there in perpetuity. But if there is the slightest deviation from that point, the system will deviate forever. Thus the point is in balance. It is an equilibrium point because it will stay where it is forever if undisturbed. However, in this case, the slightest deviation results in continued deviation. This is referred to as an unstable equilibrium point.

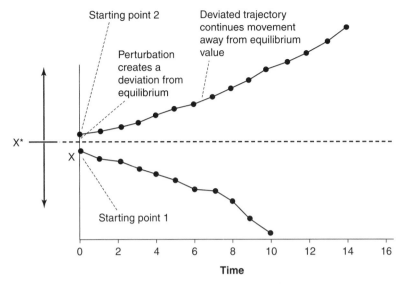

Figure 4.2. Illustration of the behavior of a point repellor.

In the past, equilibrium points have been called fixed points, singularities, and probably some other terms as well. The adjectives *stable* or *unstable* then indicate the dynamic behavior of points near that singularity (or fixed point or equilibrium point). In more recent literature, the notion of the equilibrium itself and the behavior of points near it has been termed either attractor (for a stable equilibrium point) or repellor (for an unstable equilibrium point). The terms *attractor* and *repellor* are more suitable for discussion given recent advances in our understanding of models that have the sort of complexity demanded by ecological systems. In the rest of this text, when the subject arises, the terms *attractor* and *repellor* will be used rather than *stable equilibrium* and *unstable equilibrium*.

For rapidly picturing the dynamics of a system, it is usually convenient to simply represent the attractor or repellor as a point on the line that represents the possible range of the variable in question (the ordinate in figures 4.1 and 4.2) and small arrows that indicate which direction the trajectories near the equilibrium point will go. The line is called the state space (i.e., the space, in the mathematical sense, that represents all possible values, or "states," of these variables, called state variables), and the collection of arrows is called the vector field.

A   B

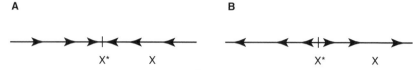

X*   X   X*   X

Figure 4.3. The state spaces of figures 4.1 (A) and 4.2 (B) (that is, the y axis turned on its side, with small vectors indicating the direction the state will change).

When the arrows point toward the equilibrium point (as in figure 4.1) it is an attractor, and when the arrows point away from the equilibrium point (as in figure 4.2) it is a repellor. Frequently, this state space is simply pictured alone, without the time dimension, as in figure 4.3. An examination of the vector field reveals whether an equilibrium point is an attractor or a repellor.

It is also popular to indicate the dynamics of a system by small physical models, as in figure 4.4. The marble on top of the mountain (figure 4.4A) illustrates the repellor (the line below it with the point and the arrows is equivalent to the ordinate of figure 4.2 turned on its side), and the marble in the bottom of the valley (figure 4.4B) illustrates the attractor. Because these attractors and repellors are all single points, they are called point attractors and point repellors.

A   B

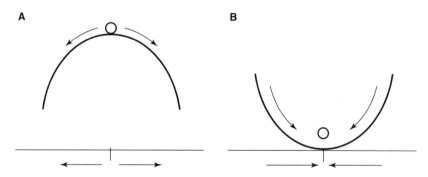

Figure 4.4. Physical models of classical attractor and repellor. (A) The marble is balanced on top of the hill at equilibrium, but the slightest deviation from that point results in continued deviation. This situation corresponds to the situation in figure 4.2. (B) The marble is located in the valley, and all deviations from that point result in a return to it. This situation corresponds to the situation in figure 4.1.

Another major category of behavior is not representable in such simple diagrams but requires a two-dimensional space. Suppose we have a beaker of water the bottom of which has the positive end of a magnet affixed to its center. We then drop a smaller magnet into the beaker with its negative pole facing downward and watch what it does as it falls through the water. In figure 4.5A the mobile magnet falls toward the magnet on the bottom. If it is placed in the water at exactly the center of the beaker, it will remain at this position (actually somewhere above this position) as it falls through the water. If it is placed somewhere deviant from this position it will fall toward the bottom magnet. Obviously, this physical model is identical to the example in figure 4.1,

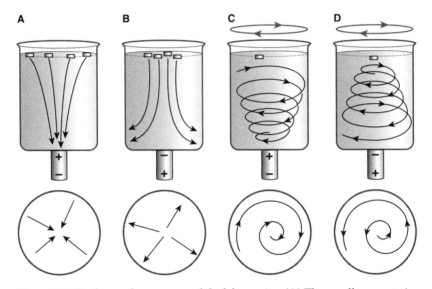

Figure 4.5. Beaker and magnet model of dynamics. (A) The small magnets in the water, with their negative poles pointed downward, are attracted to the positive pole of the magnet on the bottom of the beaker. (B) The small magnets are repelled from the negative pole on the bottom of the beaker.
(C) When the beaker is constantly rotated, the magnet undergoes a spiraling motion as it descends through the water toward the positive pole of the magnet on the bottom. (D) When the beaker is constantly rotated, the magnet undergoes a spiraling motion as it descends through the water away from the negative pole of the magnet on the bottom. The circle at the bottom of each diagram illustrates the general behavior of the small magnet as viewed from the top (or bottom) of the beaker.

except it is in three dimensions, the horizontal two dimensions of the bottom or the top of the beaker and the vertical dimension that represents time (the position from top to bottom of the beaker is proportional to the time since the small magnet was dropped into the beaker). Just as we could represent the behavior of the system on a line (the ordinates in figures 4.1 and 4.2 and the lines below the diagrams in figure 4.4), we can do so by looking at just the bottom of the beaker, as shown in the circle below the diagram of the beaker in figure 4.5. Figure 4.5A represents an attractor, and figure 4.5B represents a repellor.

With the beaker model, we can see another class of behavior that is extremely important in ecological models. Suppose the beaker is placed on a mixing table that creates a vortex in the water. The expectation is that whatever is dropped into the beaker will spiral around as it drops through the water, as indicated in figure 4.5C. However, as it spirals around, it is also attracted by the magnet on the bottom of the beaker. For obvious reasons, this attractor is referred to as an oscillatory point attractor. The behavior of the repelling magnet in swirling water (figure 4.5D) is also oscillatory, but because it is a repellor it is called an oscillatory point repellor. For each of the oscillatory points (attractor and repellor) the picture on the bottom of the beaker is a spiral (figure 4.5C,D).

Figure 4.6 illustrates how this spiraling behavior looks in a more traditional diagram of the variables over time. The two dimensions of the beaker's bottom are plotted over time to illustrate that an oscillatory attractor is the same as damped oscillations, and an oscillatory repellor is the same as expanding oscillations.

Yet another class of behavior is important in physical and biological systems. This class requires a different physical model, as illustrated in figure 4.7A. A marble on a small hill in the middle of a valley will roll down the hill but become entrapped in the valley, rolling continuously around the bottom of the valley. The sides of the valley cause any marble beginning on that surface to wind down the valley floor, again rolling around the bottom of the valley. The ultimate fate of any trajectory is either to move to the outer limits of the mountain or to wind up cycling forever in the bottom of the circular valley (presuming there is some sort of energy that keeps the system in motion). This kind of behavior is known as a periodic attractor, so named because at some time in the future the system always returns to the same position, which is to say it "periodically" returns to any given state. Note that the exam-

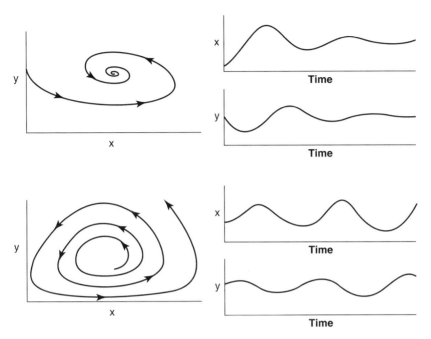

Figure 4.6. Traditional representation of an oscillatory attractor (top graphs) and an oscillatory repellor (bottom graphs).

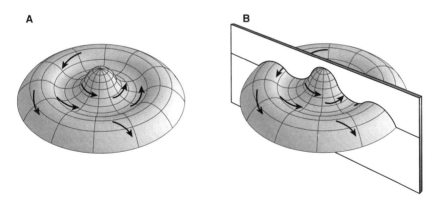

Figure 4.7. Physical model illustrating a periodic attractor (limit cycle).

ple in figure 4.7 actually includes two periodic attractors, an obvious one at the bottom of the valley and a not-so-obvious one exactly on the outer edge of the valley. That is, it is theoretically possible to have a marble cycling around the top of the outer boundaries of the valley, always exactly balanced between the force attracting it down to the valley bottom and the force attracting it down off of the side of the mountain. Obviously, this limit cycle cannot really be observed, since the marble cannot be expected to maintain exactly this balance. It is thus an unstable cycle, or a periodic repellor.

Taking a cross section of this model (figure 4.7B), we arrive at the more easily interpretable section pictured in figure 4.8 (such a section is formally a Poincaré section). As before, we can summarize the overall behavior of the system with little arrows on the line. Here we see three repellors and two attractors, although the two attractors are simply two points on the periodic attractor and the two outer repellors are simply two points on the periodic repellor. Thus, they are not really equilibrium "points" as in previous examples but rather points on an attractor (or points on a repellor).

The final type of qualitatively distinct behavior that is commonly observed in ecological models results if we assume that the bottom of the valley is perfectly flat when viewed from afar but riddled with channels when viewed close up. If the beaker model of figure 4.5 has the magnets removed from the base, or if the mountain model of figure 4.7

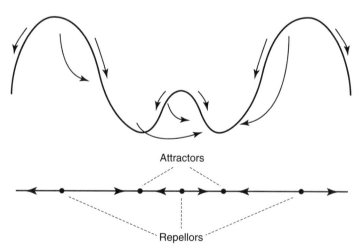

Figure 4.8. Cross section (Poincaré section) through the surface of figure 4.7, showing how the dynamics of the system can be illustrated.

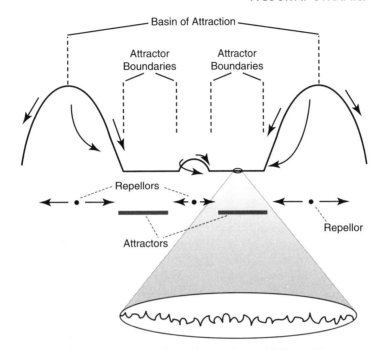

Figure 4.9. Poincaré section similar to that in figure 4.7, but with a strange attractor rather than a periodic attractor. The bottom of the valley is, when viewed from afar, perfectly flat, so there is no natural place to which the marble will be attracted. Looking at the bottom of the valley close up (the expanded oval), we see that it is riddled with many, many small valleys. Conceptually, the marble is constantly moving from one valley to another, never settling down to any one of them. So this entire flat region will attract the marble because the walls of the valley still slope downward. But once the marble reaches the floor of the valley, its motion becomes unpredictable, and it bounces from one to another of the small peaks and valleys that riddle the bottom of the larger flat valley. This is a strange attractor.

has its bottom constructed to be absolutely flat with many small channels carved in the bottom, the physical attraction (the magnet in the beaker model, the force of gravity in the mountain model) has been removed, and we theoretically expect the system to move around in this space, constrained to be sure, but without a central tendency to which the motion tends within that space, as suggested by the model in figure 4.9. As before, there are three repellors, but this time there are no attractors, at least not of the sort in the previous examples. Yet, the entire bottom of the valley will certainly attract the marble, and in this intu-

itive sense is also an attractor. But here we have an attractor that is neither a point nor a cycle; it is, rather, an area or a region. Once the marble reaches that region, it will bounce around in those small channels,
never remaining in one for any length of time, always being, in a sense,
on the top of a small ridge. Being a region that attracts all trajectories
yet has no central tendency within it (no point attractor) is thought to
be rather strange. For this reason, it is referred to as a strange attractor,
and the behavior of a system within it is referred to as chaotic.

Obviously, there is a qualitative difference between this type of attractor and those discussed previously. There is no particular point to
which the system ultimately tends, but rather an area to which it
tends. Furthermore, from a practical standpoint we cannot be interested in the final state of the system because it has multiple final states
(in the sense of a single point). The concern really ought to be with the
range within which the system will ultimately be found, as discussed
below.

The classes of behavior illustrated by the simple hills (figures 4.4
and 4.7) and the beakers (figure 4.5) are the classical behaviors usually analyzed by engineers. Certainly, they are important points of
departure for analyzing ecological systems. However, there are other
kinds of behavior, most importantly as illustrated in figure 4.9, in
which the focus is on the range of expected values and the persistent
changes through time within that range. In summary, the first class
of behaviors may be referred to as point attractors and point repellors (figures 4.1, 4.2, and 4.5), the second class as periodic attractors
and repellors (figures 4.7 and 4.8), and the third class as strange attractors and repellors (figure 4.9). Attractors (and repellors) can be
thought of as falling on a gradient going from simple point attractors (or stable equilibrium, stable node, stable fixed point—all syn-
onyms—as in figure 4.3A), to oscillatory attractors (or stable focus—
still a point attractor—as in figure 4.5C), to periodic attractors (also
called limit cycles—as in figure 4.7), to strange attractors (or chaotic
attractors, figure 4.9; the subject of chaos will be discussed later in
this chapter).

One's interest in analyzing a system depends on the nature of the
equilibrium state. If there is a point equilibrium, for example, one
wants to find the exact position of the point and determine whether it
is an attractor or a repellor. This is the focus of the classical engineering
sciences. But if there is a strange attractor, one wants to locate its

boundaries and discover other details about its "morphology," as discussed later.

One further concept is especially important when dealing with strange attractors. The "basin of attraction" is the collection of the values of the state variables from which all trajectories eventually wind up exactly on the attractor. In figure 4.8, for example, the tops of the two largest hills represent the outer edges of the basin of attraction for the limit cycle attractor at the bottom of the valley, and the small hill in the middle represents the inner edge of that basin. The edges of a basin of attraction are always repellors, as is evident in figure 4.8. The edges of the basin of attraction are not the same as the boundaries of a strange attractor. The latter refer to the outer limits that the attractor itself can realize, the former to all possible states that eventually reach the attractor. Formally speaking, the attractor (and its boundaries) is a subset of the basin of attraction, but not the reverse (see, for example, figure 4.9).

For the most part, classical ecological theory has dealt mainly with point attractors and to some extent with periodic attractors. Only with the advent of nonlinear dynamics as a theoretical science has there been a realization that an alternative type of equilibrium and stability actually exists: that is, the strange attractor. Such attractors have become more common in the literature as old models are analyzed more completely and especially as new model situations are explored. These attractors have also received considerable attention simply because they are sometimes called chaos, or chaotic attractors. This unfortunate choice of terminology will be further discussed later in this chapter. For now, suffice it to say that a strange attractor, because of its so-called chaotic motion, is unpredictable in a very special technical sense. This fact has caused considerable unnecessary consternation for scientists who seek to predict natural phenomena and has led to a small cottage industry of attempts to show that particular data sets do or do not represent true chaotic behavior. However, the importance of the issue lies not with the distinction between chaos and nonchaos but with the distinctions among point, periodic, and strange attractors, which are three positions on a continuum from point to strange, as discussed above. In the case of point attractors, we are concerned with the location of the equilibrium and its stability properties. In the case of strange attractors, we are concerned with their boundaries and qualitative behaviors, their morphology.

Eigenvalues: A Key Concept in Dynamic Analysis

Consider a simple point attractor in one dimension. As above, we can represent its qualitative dynamics by drawing the state space (a line) and indicating where the point is in that space (on that line) and then adding the vector field (small arrows indicating the direction and rate of change), as presented in figure 4.10. If we now rotate the vectors 90 degrees, either upward to illustrate an increasing vector or downward to illustrate a decreasing vector, we get the picture shown in figure 4.11. The slope of a line connecting the tips of the rotated arrows is called the eigenvalue. With this formulation (figure 4.11) it is evident what the eigenvalue means; it is the rate at which the system approaches a point attractor, that is, if the system actually has a point attractor.

Recall the exponential equation from chapter 1,

$$\frac{dN}{dt} = rN$$

This is a system of one dimension (a single variable), and thus its state space and dynamics are as in figure 4.10. The equilibrium point is at $N = 0$. Suppose $r < 0$, which is to say, a declining population, one that will eventually go locally extinct. A plot of dN/dt versus N will look something like figure 4.11 (since the vectors represent dN/dt at particular values of N, rotating them 90 degrees is the same as plotting them on the y axis), with the slope $= r$. Thus we see that, in this example, the eigenvalue of the attractor is r.

If the population is growing, the result is qualitatively different in that the arrows will all be pointing away from the equilibrium point;

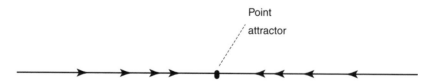

Figure 4.10. State space for a one-dimensional (one-variable) model, illustrating a single point attractor and its vector field (the collection of arrows indicating the dynamics of the system).

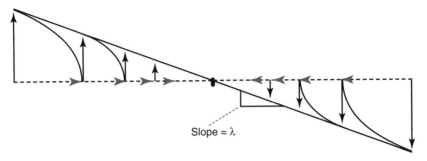

Figure 4.11. Rotating the vectors of the example from figure 4.10. The vectors to the right of the point attractor are decreasing; thus we rotate them downward (decreasing). The vectors to the left of the point attractor are increasing; thus we rotate them upward (increasing). Then we connect the arrowheads with a line. The slope of the line is the eigenvalue of the point attractor.

that is, the point is a repellor (the equilibrium point is still 0). Thus, the arrows to the right of the point will be rotated clockwise while the arrows to the left of the point will be rotated counterclockwise (the opposite of figure 4.11). The line connecting the arrowheads will thus have a positive slope.

Suppose we have a population growing according to the logistic equation. Its dynamics (again in one dimension) will look something like that pictured in figure 4.12. Here we have two equilibrium points, one an attractor at the carrying capacity and one a repellor at the value of 0. If we now rotate the arrows, as before, we obtain the graph shown in figure 4.13. Here, there is no simple slope to the line; in the neighborhood of each of the equilibrium points, however, we can approximate the curve with a straight line, and the slope of that straight line is the eigenvalue associated with the equilibrium point. For example, con-

Figure 4.12. State space for a one-dimensional model based on the logistic equation. There are two equilibrium points, one an attractor (K) the other a repellor (0).

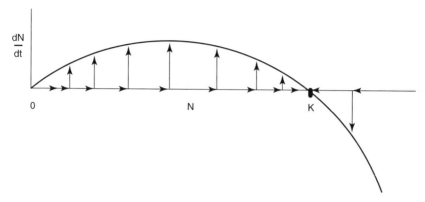

Figure 4.13. Dynamics of the logistic equation in one dimension, with the changes in the derivative graphed as the ordinate.

sider the equilibrium point at the carrying capacity (*K*). The slope of the curve at that point is simply the derivative with respect to *N* of the derivative with respect to time, evaluated at *K*, or,

$$\frac{d\left(\dfrac{dN}{dt}\right)}{dN} = r - \frac{2rN}{K}$$

and substituting *K* for *N* (since the slope is the derivative evaluated at *K*), we obtain,

$$\frac{d\left(\dfrac{dN}{dt}\right)}{dN}\Bigg|_{N=K} = r - \frac{2rK}{K} = r - 2r = -r$$

which tells us first, the point *K* is an attractor (since the eigenvalue, −*r*, is negative), and second, the rate of approach to that equilibrium will be −*r*.

A similar procedure applied at the other equilibrium point (0) gives an eigenvalue of *r*, showing that it is a repellor (because *r* is positive), and the rate of deviation from it is *r*.

Note that the eigenvalues computed for the projection matrices of chapter 3 have precisely the same qualitative meaning as those in this chapter. However, we discussed the computation of eigenvalues only for a matrix with constant values, in which case the population was always an exponential population, with a single equilibrium point at zero. If the poulation was growing, its largest eigenvalue was positive and it was growing at a rate equal to the value of that eigenvalue. A negative dominant (largest) eigenvalue indicated, as it does here, that the equilibrium point is an attractor, which means that the population is declining, and the rate of that decline is the value of that eigenvalue. So we see that the dominant eigenvalue of a projection matrix (without density dependence) is precisely the same as the eigenvalue of the exponential equation, r.

In two or more dimensions (i.e., when we have predation, competition, or mutualism occurring), the situation is a bit more complicated. In two dimensions, the state space is the plane, and we must examine the dynamics of the system in that plane in a third dimension. We have already looked at this issue in a very qualitative way in figure 4.5A,B (representing the stationary beaker), where the fixed magnet either attracted or repelled the falling magnet. We now examine the two-dimensional case in more detail. Consider the physical model in figure 4.14. A marble rolling on this surface will eventually wind up at the point where the two folds intersect, but there will be a bias in that most of the time it will roll down along the fold labeled A. The marble could roll directly down fold B and arrive at the equilibrium point, but its doing so would be very unlikely because that fold is a knife edge on which the marble would have to balance as it rolled down.

For heuristic purposes, it makes sense to ask what would happen if the marble began exactly on the fold A. Now we can represent the system in a single dimension, a dimension along the fold A, and look at the dynamics along this fold as if it were a one-dimensional system. Thus, the analysis reverts exactly to the one-dimensional analysis we did in figures 4.10 through 4.13. And indeed the rate of change of the rate of change along this fold is an eigenvalue.

But there is still the theoretical possibility that the marble will balance on fold B and, like a tightrope walker, roll down, precariously balanced, until reaching the equilibrium point. As unlikely as that possibility may seem, we can still analyze it mathematically, using the graphical method used in figures 4.10 through 4.13. Again, we come up with a measure of the rate of change of the rate of change as we approach the equilibrium point, and that rate is an eigenvalue. Thus, we

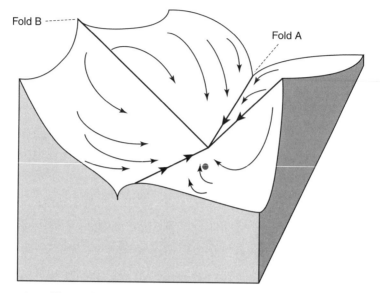

Figure 4.14. Physical model of the dynamics of a point attractor in two dimensions.

see how, when we have two dimensions, we have two eigenvalues. Indeed, it is the case that there will always be as many eigenvalues as there are dimensions in the system.

Here we see the significance of the "dominant" eigenvalue. It is the value of the rate of change of the rate of change along the dominant fold (fold A in figure 4.14): that is, the rate at which the system will approach the equilibrium point as it gets close to it. There will always be one collection of points (a "fold") along which the marble will eventually tend, and that collection of points defines the one-dimensional system that is used to calculate the dominant eigenvalue (see figures 4.10 and 4.11).

In figure 4.15, the three possible configurations in two dimensions are illustrated, along with the eigenvalue states. Clearly, an examination of the signs of the eigenvalues provides a definitive statement as to which of the situations exists. Two positive eigenvalues indicate a simple point repellor; two negative eigenvalues indicate a simple point attractor; and a positive and a negative eigenvalue indicate a different sort of repellor. Note that the point is approached from some lines and

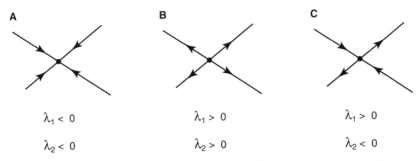

Figure 4.15. Conditions of eigenvalues for the three most common qualitatively distinct arrangements in two dimensions. (A) Point attractor. (B) Point repellor. (C) Saddle point repellor.

repelled along other lines, much as a marble would be when rolling along the surface of a saddle. For this reason, this sort of equilibrium is referred to as a saddle point repellor.

In all of the above, the presentation is largely graphical and heuristic. In reality, for a given model there are simple recipes one uses for finding the eigenvalues of a system at a point (indeed, in the contemporary world making a few keystrokes or pointing and clicking on the "find eigenvalue" button is usually the way to do it). Frequently, the eigenvalues come out to be simple real numbers and one merely has to compare them with zero to determine the qualitative nature of the point. But sometimes they come out as complex numbers, that is,

$$\lambda = r + ci$$

where i is the square root of negative 1. Thus, there is a real part (r) and a complex part (c). There is no convenient way of explaining exactly why, but the fact is that oscillatory systems (e.g., the swirling beaker model of figure 4.5) have eigenvalues with non-zero complex parts. The parallel graphs for the ones already made in figure 4.15 are shown in figure 4.16 for oscillatory point attractors and repellors. It is a simple rule that if the imaginary parts of the eigenvalues are non-zero the system is oscillatory, and the oscillations wind down to the equilibrium point if the real values are negative and wind away from the point if the real values are positive.

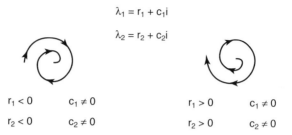

$$\lambda_1 = r_1 + c_1 i$$

$$\lambda_2 = r_2 + c_2 i$$

| $r_1 < 0$ | $c_1 \neq 0$ | | $r_1 > 0$ | $c_1 \neq 0$ |
| $r_2 < 0$ | $c_2 \neq 0$ | | $r_2 > 0$ | $c_2 \neq 0$ |

Figure 4.16. Conditions of eigenvalues for the two most common qualitatively distinct arrangements in two dimensions when the eigenvalues have non-zero complex parts.

Basic Concepts of Equilibrium and Stability in One-Dimensional Maps

Let us suppose that a reasonable model of population dynamics is a mapping from one time period to the next, which is to say, the population density at this point in time is some multiple of what it was the previous time period. That is, if N_t is the population density at time t, we have, for example,

$$N_{t+1} = aN_t \tag{1}$$

This is an alternative way of expressing the exponential growth of a population. The relationship between this form and the differential equation of chapter 1 is as follows. From the equation,

$$\frac{dN}{dt} = rN$$

integrate, to obtain,

$$N_t = N_0 e^{rt} \tag{2}$$

which means we can also write,

$$N_{t+1} = N_0 e^{r(t+1)} = N_0 e^{rt} e^r$$

and substituting from equation 2, we have,

$$N_{t+1} = N_t \, e^r$$

and letting $a = e^r$ we substitute to obtain equation 1, making it obvious that the one-dimensional map is exactly equivalent to the more traditional differential equation.

The One-Dimensional Map

The one-dimensional map is a convenient modeling technique especially because of its obvious graphical interpretation: it is possible to gain a rapid idea of the dynamic behavior of a model simply by glancing at a graph. The map applies to systems that can be represented as the projection of a variable from one time unit to the next. First construct a graph of the population density in year t versus the population density in year $t+1$. Suppose, for example that the population density, beginning in year 1997 is 20, and in subsequent years it is 40, then 80, then 100, then 110. That is, $N_{1997} = 20$, $N_{1998} = 40$, $N_{1999} = 80$, and so on. To graph the numbers in the style of a one-dimensional graph (one-dimensional because only one dynamic state variable is under consideration), we graph first 20 on the abscissa and 40 on the ordinate, then 40 on the abscissa and 80 on the ordinate, then 80 on the abscissa and 100 on the ordinate. In doing so, we are essentially making a graphical form of the number series 20, 40, 80, 100, 110. We know that the number 20 projects into 40, and drawing an arrow from 20 on the abscissa to its intersection with the value of 40 on the ordinate is simply a graphical statement of this fact (that 20 projects into 40). We now wish to project from 40, in which case we draw a similar arrow from 40 on the abscissa to where it intersects the value of 80 on the ordinate. The first projection (from 20) yielded 40, and we sought to initiate the second projection from this value of 40. And this is a general rule. The next projection always begins where the previous projection left off. How can we know where that initiation is? We can search for the ordinate value on the abscissa (i.e., from the first projection from 20 to 40, we search for 40 on the abscissa so as to make the second projection). But that search is made graphically much simpler if we draw a reference line, beginning at zero for both abscissa and ordinate and rising at a 45 degree angle to the axes. We can now take the original projection and reflect it back to the 45 degree line. It is a graphical technique for locating the projected value on the abscissa so that it can be projected into the next time period. This whole example is illustrated in figure 4.17.

This example can now be generalized. Instead of specific numbers, we may write a general rule of projection. For example, N at time t will become $N + 5$ at time $t + 1$ ($N_{t+1} = N_t + 5$); or N next year will be twice the value of N this year ($N_{t+1} = 2N_t$); or simply N next year depends on the value of N this year; that is, N next year is a function of N this year ($N_{t+1} = f[N_t]$). Although it is frequently possible to state the exact relationship between N this year and N next year, in the absence of that knowledge it is also useful simply to be able to draw the general shape

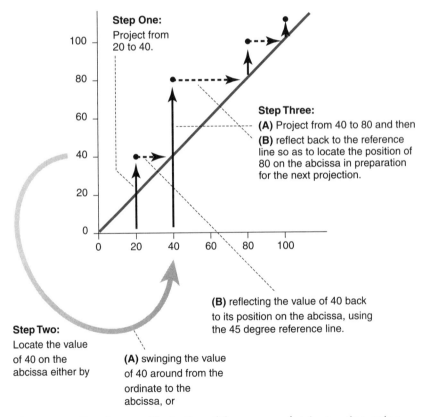

Figure 4.17. Step-by-step illustration of the process of stair-stepping, using numerical values, for a one-dimensional map.

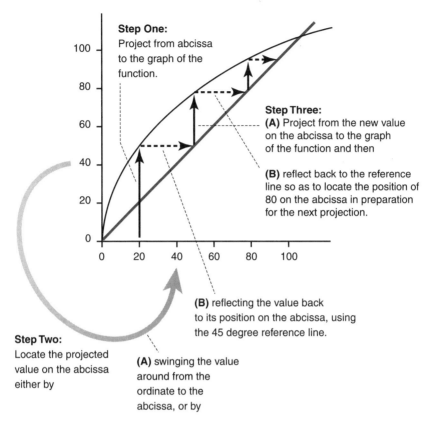

Step One:
Project from abcissa
to the graph of the
function.

Step Three:
(A) Project from the new value
on the abcissa to the graph
of the function and then

(B) reflect back to the reference
line so as to locate the position of
80 on the abcissa in preparation
for the next projection.

(B) reflecting the value back
to its position on the abcissa, using
the 45 degree reference line.

Step Two:
Locate the projected
value on the abcissa
either by

(A) swinging the value
around from the
ordinate to the
abcissa, or by

Figure 4.18. Step-by-step illustration of stair-stepping using a function.

of f, which is frequently possible from only qualitative knowledge of
how the system behaves. But the stair-stepping procedure is still the
same. Instead of projecting from 20 to 40 (as in the above example),
project from an arbitrary starting point on the abscissa to the graph of
the function. Then locate that projected value on the abscissa by reflecting it back to the 45 degree reference line and project it to the graph of
the function again. This process is illustrated in figure 4.18.

The general rule is, project to the function, reflect to the reference
line, project to the function, reflect to the reference line, and so on. After a short practice session, the qualitative dynamics of almost any one-
dimensional map can be visualized with a simple glance at the graph.

In figure 4.19, equation 1 is graphed along with the classical stair-
stepping technique that can be used to quickly visualize the dynamics

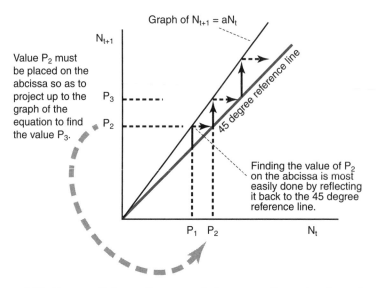

Graph of $N_{t+1} = aN_t$

N_{t+1}

Value P_2 must be placed on the abcissa so as to project up to the graph of the equation to find the value P_3.

P_3

P_2

45 degree reference line

Finding the value of P_2 on the abcissa is most easily done by reflecting it back to the 45 degree reference line.

P_1 P_2 N_t

Figure 4.19. Exponential equation presented as a one-dimensional graph. The qualitative dynamics of such a system are easily visualized with the stair-stepping technique. Beginning at point P_1, go up to the graph of the equation to reach P_2 on the ordinate. The ordinate value P_2 must then be positioned on the abcissa, which is most easily done by reflecting it to the 45 degree reference (dashed arrow), which indicates its position on the abcissa. From P_2 on the abcissa, go up to P_3 and repeat the process (see figures 4.17 and 4.18).

of the system. Where the graph of the equation crosses the 45 degree line, an equilibrium point exists. For equation 1 (the well-known exponential equation) that equilibrium is at zero. If $a > 1.0$, the particular nature of that equilibrium is unstable, because any value of N close to the point (i.e., the equilibrium $N = 0$) will deviate away from it. If the point were set at exactly $N = 0$, a glance at equation 1 shows that it will stay there forever. But if there is the slightest increase from zero (e.g., $N = 0.0000001$), the population will grow and thus deviate from the equilibrium point.

Suppose now that at each time interval a constant number of individuals migrates into the population. Suppose that number is m, thus transforming equation 1 into,

$$N_{t+1} = aN_t + m \qquad (3)$$

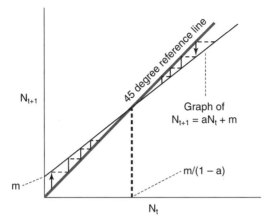

Figure 4.20. Graph of equation 3, illustrating a point attractor. The stair-stepping technique is the same as in figures 4.17 and 4.18. Any point initiating either above or below the attractor (where the graph of the function crosses the 45 degree reference line) eventually approaches that point. It is thus a point attractor, since any deviation from it will automatically revert to it.

A graph of equation 3 is presented in figure 4.20 (assuming $a < 1.0$). Once again, the point at which the graph of the equation crosses the 45 degree line is an equilibrium point: setting N at exactly that point, which in this case is $m/(1-a)$, results in the same value of N for every future time period. This time, however, the equilibrium is a stable one, as illustrated in figure 4.10. Whatever the initial population size, the tendency will be to return to the value of $m/(1-a)$, the equilibrium state, which is thus an attractor.

Assume now that, instead of regular immigrants into the population, a human population of hunters harvests a particular number of individuals from the population each time unit. Thus, a constant number, p, of individual prey organisms will be taken out of the population each time unit, and the appropriate equation is,

$$N_{t+1} = aN_t - p \qquad (4)$$

which is graphed in figure 4.21 (with the assumption that $a > 1.0$). Note that the equilibrium point is $p/(a-1)$, and it is an unstable one, making the point a repellor, just as the zero equilibrium point was unstable for the original exponential equation (equation 4).

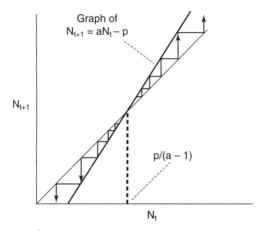

Figure 4.21. Graph of equation 4, illustrating an unstable equilibrium. The stair-stepping technique is the same as in figures 4.17, 4.18, and 4.19. Any point deviating only slightly from the equilibrium will continue deviating. It is thus a point repellor, since any deviation from it will continue deviating (it "repels" all values).

These two modifications to the basic exponential equation have both been linear. However, most ecological processes of interest are known to be nonlinear, so it makes sense to modify equation 1 in a nonlinear fashion. The most elementary nonlinearity would be to assume that the parameter that multiplies the variable of interest (i.e., a, in equations 1, 3, and 4) is itself a function of the variable. If we assume that the parameter a is a decreasing function of N (the growth of the population depends on its density; recall density dependence from chapter 1), and furthermore that the exact function is $a = r - rN$, the exponential equation (equation 1) becomes,

$$N_{t+1} = rN_t(1 - N_t) \tag{5}$$

In figure 4.22, equation 5 is graphed for two values of r. From the simple stair-stepping graphical technique, it is obvious that both cases pictured are oscillatory. That is, at successive intervals the population alternately increases and decreases, as illustrated in the diagrams beneath the stair-stepped graphs. The difference between figure 4.22A and 4.22B is the difference between an oscillatory attractor (figure 4.22A) and an oscillatory repellor (figure 4.22B).

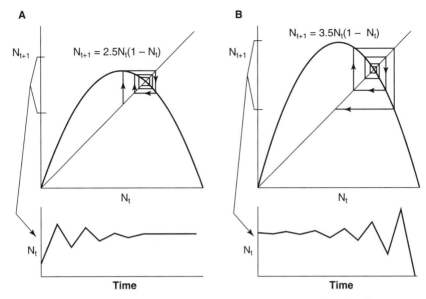

Figure 4.22. Graphs of equation 5. (A) Oscillatory attractor. (B) Oscillatory repellor. The bottom graphs illustrate the behavior of the variable through time.

Figures 4.20, 4.21, and 4.22 summarize the classical notions of equilibrium and stability. A single point is the equilibrium point, and it may be an attractor (stable) or a repellor (unstable), oscillatory or nonoscillatory. Because we are now dealing with discrete space rather than continuous space, the eigenvalue rules as elucidated in the previous section do not directly apply. Indeed, since the eigenvalue is the slope of the function as it crosses the 45 degree line, the above development shows that the eigenvalue greater or less than 1 or −1 stipulates the qualitative nature of the equilibrium point, as illustrated in figure 4.23. If the eigenvalue is >1, the system is a point repellor. If the eigenvalue is <1 but >0, the system is a point attractor. If the eigenvalue is <0 but >−1, the system is an oscillatory attractor. If the eigenvalue is <−1, the system is an oscillatory repellor.

If the system generates a strange attractor (or for that matter a permanent cycle), ideas of point attractors and repellors are useless, despite the fact that point repellors are always contained within strange attractors. For example, in figure 4.24 three cases are illustrated in which the critical equilibrium point (where the graph of the equation crosses the 45 degree line) is a repellor. However, knowing that the

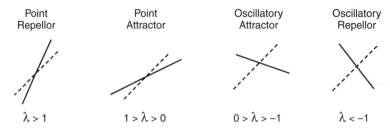

Figure 4.23. Eigenvalue values for the various forms of stability in a one-dimensional map. The dotted line is the 45 degree reference line, and the solid line is the function (illustrating that part of the function near its crossing with the 45 degree line).

equilibrium point is a repellor provides us with scant information on what is interesting about the behavior of the system. Indeed, what is important, in this case, is the distinction between cases A and B on the one hand and C on the other. Cases A and B will persist indefinitely (i.e., they are sustainable), whereas in case C, the variable is extinguished from the system (the population goes locally extinct). If this were, for example, the population of an introduced natural enemy in an agroecosystem, we care little that the population is theoretically unstable (its equilibrium point is unstable). Rather, we are concerned with whether the new natural enemy will persist in the environment: that is, whether we are dealing with, on the one hand figure 4.24A or 4.24B, or, on the other hand, figure 4.24C. The interest here is not in the equilibrium point itself but in the limits, or boundaries, of the system. These boundaries are illustrated by dashed lines intersecting the two axes in figure 4.24A,B.

A further word is in order regarding the difference between the patterns in figure 4.24A and 4.24B. Figure 4.24A is classically known as an n-point cycle (the particular value of n in figure 4.24A is 2, because there are two actual values of N that repeat themselves forever, as indicated by the dashed lines crossing the axes). It is the one-dimensional equivalent of the classical limit cycle in the context of differential equations. Figure 4.24B represents an example of chaos, better termed a strange attractor. The equilibrium point is unstable, yet there is no single n-point cycle, and trajectories essentially move in a totally unpredictable direction, giving rise to the popular appellation *chaos*. An enormous literature is devoted to the analysis of and significance of the

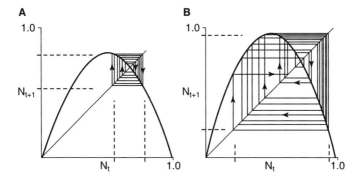

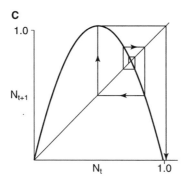

Figure 4.24. Graphs of equation 5. (A) A two-point periodic attractor ($r=3.1$). (B) A strange attractor ($r=3.8$). (C) An oscillatory repellor, leading to extinction of the population ($r=4.2$). In cases A and B the population has a repellor (is unstable) where the function graph crosses the 45 degree line, but in both cases the repellor is constrained by dynamic boundaries. Cases A and B are thus referred to as regionally stable.

difference between the pattern in figure 4.24A and that in 4.24B (Hastings et al. 1993; Elner and Turchin 1995). To some extent, that literature has been misdirected. The key question seems to have become whether ecological systems are chaotic. But for understanding ecological systems it is not clear how knowing whether a system is chaotic or has a 50-point cycle will make much of a difference! True, a chaotic system is theoretically unpredictable, but in practical fact an n-point cycle is almost as unpredictable if n is large. On the other hand, in both figure 4.24A and 4.24B there is a clear structure to the trajectories. Both are fundamentally oscillatory, even though the peaks of the oscillations do not repeat themselves exactly each year in the case of figure 4.24B and, most important, both have limits that they will never transcend (in a strictly deterministic world).

These limits are essentially identical to what Lewontin (1969) has referred to as dynamic boundedness. They define a section of the state space into which all nearby trajectories will eventually enter but which no trajectory can ever exit. Because all nearby trajectories must enter this space, the space itself is called an attractor, even though the equilibrium point within that space is unstable. Whether an attractor is strange or periodic will not be an important focus of this chapter. The significant practical features for understanding ecosystem dynamics are the location of the boundaries and the qualitative structure of the dynamics, for both periodic and strange attractors (and repellors). Thus the important question to be asked of an unstable point is whether the nonlinearities of the system create boundaries around that point, thus making it either a periodic or a strange attractor. If boundaries are not created around the point, the system will extinguish itself.

Stability and Equilibrium in the Logistic Map

The logistic map (as equation 5 is usually called) can be used to illustrate these and other simple ideas in a straightforward manner. The equilibrium point is,

$$N^* = (r-1)/r$$

(there is another, trivial, equilibrium point at $N^* = 0$). This means that r must be greater than 1.0 to have a positive equilibrium point, K, and the equilibrium will be stable and nonoscillatory whenever the derivative of the function, evaluated at the equilibrium point, is greater than zero (these conditions, and the ones that follow, should be clear after a detailed examination of the graph of equation 5). That is, differentiating equation 5, we obtain,

$$(dN_{t+1}/dN_t) = r - 2rN_t$$

which, when evaluated at the equilibrium point (i.e., substitute $N_t = [r-1]/r$) is,

$$(dN_{t+1}/dN_t) = r - 2r[(r-1)/r] = r - 2r + 2 = 2 - r$$

in which case we see that as long as $r < 2$ the equilibrium point will be a simple nonoscillatory attractor (note that the case of an unstable nonoscillatory point, as shown in figure 4.21, is not possible with this model). When $r > 2$, the system will be oscillatory. It will have an attractor if the derivative of the function evaluated at the equilibrium point is greater than −1. Thus,

$$2 - r > -1$$

or,

$$2 < r < 3$$

indicates an oscillatory attractor, and

$$2 - r < -1$$

or

$$r > 3$$

indicates an oscillatory repellor. Note that the value of the eigenvalue is $2 - r$, making these observations consistent with the eigenvalue conditions of figure 4.23.

The existence of an oscillatory repellor leads to the further question of how to distinguish between persistence (cases A and B of figure 4.24) and extinction (case C of figure 4.24). Extinction will occur only if the projection from the maximum value of the map falls on the x axis at a point greater than the intersection of the function (see figure 4.25).

From an examination of the original equation (equation 5), we see that the function intersects the abscissa at 0 and 1. Thus, extinction of the system will occur when the projection from the peak of the map touches the x axis at a point > 1 (see figure 4.25A). The peak occurs at $N_t = 0.5$, so its projection (from equation 2) will be,

$$N_{t+1} = r(0.5)(1 - 0.5) = 0.25r$$

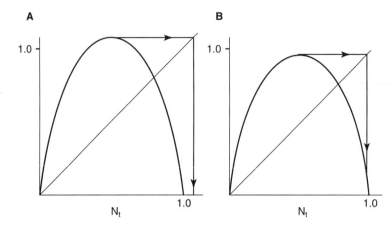

A   **B**

N_t N_t

Figure 4.25. The difference between stability and instability in the regional sense. (A) Regionally unstable, the population will go extinct. (B) Regionally stable, the population will persist (albeit in a chaotic state). In both cases, the equilibrium point is unstable in the neighborhood sense.

and the condition for extinction is thus,

$$0.25r > 1$$

$$r > 4$$

So we see the system will have either a periodic or a strange attractor whenever

$$3 < r < 4$$

We could thus say that the system will be sustainable as long as r is between 1 and 4, even though the classic conditions for stability fail for $r > 3$. Although the specific development in this chapter is associated with the density of a single population, the same dynamic rules apply if the state variable is some other interesting variable. For example, N might be the yearly production of manure from a dairy farm, or the soil organic matter in a forest system, and so forth. If we presume that equation 5 represents the system, we can unambiguously define sustainability as $1 < r < 4$. The trick, of course, is that equation 5 is normally too simple to accurately represent anything as complicated as organic matter or manure (or even population density), and we use it here for didactic purposes only.

The upper and lower boundaries of the system are easily calculated. The upper limit is simply the peak of the function,

$$N_{t+1} = r(0.5)(1-0.5) = 0.25r$$

and the lower limit is its projection,

$$N_{t+1} = r(0.25r)(1-0.25r) = 0.25r^2 - 0.0625r^3$$

Again, depending on the context, such boundaries may be of tremendous interest. For example, if N is the population density of a pest insect and the damage threshold is known (say it is D), the population will never be a pest if $0.25r$ (the upper threshold) is less than D. Thus $r < 4D$ exactly stipulates the conditions under which this population will be a sometime pest.

Basins of Attraction in the Logistic Map

For most simple models of ecological processes it has been possible to analyze the equilibrium points and leave it at that. Most ecologists now admit that more complicated models are necessary to reflect even the simplest ecological phenomena. With even slightly more complex models, we face a situation in which alternative equilibria exist in the same model. For example, if we combine the ecological principle that led to equation 4 with the one that led to equation 5, we obtain

$$N_{t+1} = rN_t(1-N_t) - p \qquad \text{for } N_t > 0 \qquad (6a)$$

where r is again the rate of population increase, and p is the number of individuals removed from the population each time unit by a constant predator. To make equation 6 relevant to ecological processes, we restrict its application to $N_t > 0$ and add the equation

$$N_{t+1} = 0 \qquad \text{for } N \leq 0 \qquad (6b)$$

This condition simply acknowledges that there can be no values of N less than 0 (assuming the running example of N signifying the population density of a pest insect; other variables may take on negative val-

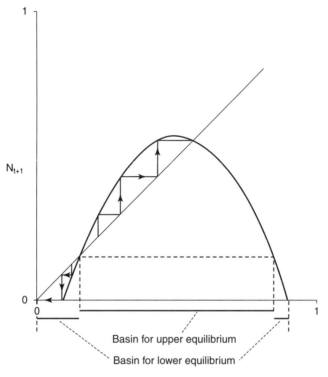

Figure 4.26. Graph of equation 6, illustrating the two basins of attraction for the two point attractors.

ues, in which case the special condition for $N < 0$ would not be necessary). Equation 6 is graphed in figure 4.26. There are three equilibrium points, given as

$$N^* = 0 \text{ (from equation 6b)} \tag{7a}$$

$$N^* = [(r-1)/2r] + \{[(r-1)/2r]^2 - p/r\}^{1/2} \tag{7b}$$

$$N^* = [(r-1)/2r] - \{[(r-1)/2r]^2 - p/r\}^{1/2} \tag{7c}$$

Inspection of figure 4.26 shows that the central equilibrium point is a repellor; the lower one (at $N = 0$) is an attractor, as is the upper one. Knowing the location of the equilibrium points is important, but there is another feature of figure 4.26 that is of importance to understanding the population dynamics. Any value of N near to but greater than the repellor will eventually approach the upper attractor, while any point less than the repellor will eventually approach the lower attractor. The repellor thus separates the state space (all possible values of N) into those values that approach the upper and those values that approach the lower attractor (this statement is only approximately true, as discussed in the next paragraph). The unstable point in this context is referred to as a separatrix, and the collection of points on either side of it, as a basin of attraction. The basin of attraction refers to the section of the relevant space in which all trajectories approach a given attractor. This concept—which of the initial values will eventually reside in particular locations—may turn out to be far more important for analyzing ecosystems than the traditional question of the exact location of the equilibrium point and whether it is stable.

This issue is a bit more complicated in the case of the logistic map, as illustrated in figure 4.26. Very high values of N, because of the strong density dependence of the logistic model, will be projected to values of N less than the separatrix. Thus, there is a section of the lower equilibrium's basin of attraction that exists at very high values of N, in addition to the obvious one that exists at lower values of N.

Structural Stability

A notion of stability totally distinct from that discussed previously may arise when parameters undergo change. That is, in all the above examples, the state variable (X_t or Y_t or N_t), the one that is dynamic (i.e., that varies through time), is clearly distinguished from the parameters, which are not dynamic (i.e., do not vary through time). For example, in equation 6a, N_t is the state variable, while r and p are parameters. For purposes of analysis we presume that N_t varies through time but r and p do not.

A different sort of analysis emerges when we ask what happens when r and p themselves vary along a distinct time frame. For example, N_t may vary in ecological time while r changes slowly through evolution and p may change as the resident predator population slowly increases or decreases. It is of great interest to examine what happens to the general results as the parameters change and what this result has to do with stability. In this context, when we speak of "state space" or "parameter space" we are speaking of two different things.

State space is represented as a graph of the potential value (or values) of the state variable (or variables), whereas parameter space is represented as a graph of the potential values of all the parameters in the model.

Consider the case of a nonreproductive population receiving migrants in a density-dependent fashion. That is, suppose the rate of migration into the population is f, but f itself is a negative function of population density (the migrants have the ability to sense when the population is overcrowded, for example, and tend to avoid an overcrowded situation). This situation could be modeled with the simple equation

$$N_{t+1} = r(1 - N_t) \tag{8}$$

As illustrated in figure 4.27, if the value of r is greater than 1.0, the equilibrium point is oscillatory and a repellor (figure 4.27A). If the value of r is less than 1.0, the equilibrium point is oscillatory and an attractor (figure 4.27C). The question then arises, What if the value of r is exactly 1.0? Mathematically, such a situation presents precisely the behavior one would expect, oscillatory and neither an attractor nor a repellor (figure 4.27B). Although it may not seem particularly important that the population permanently oscillates between two particular values, the form of oscillation is particularly unusual. At every second time projection, the population will return to exactly what it had been before, no matter where it started. For example, if we begin with $N = 0.3$, the next value will be 0.7, (see equation 8) and the next value 0.3 again; if we begin with 0.2, the next value will be 0.8 and the following one 0.2 again. That is, the population will oscillate with a cycle that is two time periods in length, but the exact values of the cycle depend on the starting point. Although apparently biologically uninteresting, this situation is key for conceptualizing the idea of structural stability. The behavior is illustrated in figure 4.27B, in which we see that the oscillations are neither stable nor unstable. This state is dependent on the assumption that $r = 1.0$. If $r = 0.9999$ (say), the population will no longer continue returning to the point at which it started but, rather, will slowly converge on the value of about 0.67 (Figure 4.27C; actually, the value is about 0.66665555). That is, the qualitative behavior of the system changed dramatically when the value of r was changed only slightly, from a system that was dependent on the start-

$$N_{t+1} = r(1 - N_t)$$

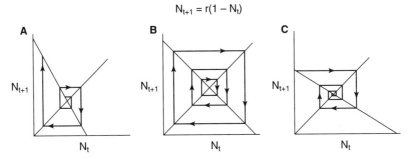

Figure 4.27. Illustration of structurally unstable parameter configuration for equation 8: $N_{t+1} = r(1 - N_t)$. (A) A point repellor resulting from a slightly larger value of r (i.e., $r > 1$). (B) A neutrally stable situation in which the initiation point is forever repeated every other time unit (i.e., $r = 1$). (C) A point attractor resulting from a slightly smaller value of r (i.e., $r < 1$).

ing point, ever cycling back to the same point, to a system that converged on a single equilibrium point, an attractor. Similarly, if $r = 1.00001$, the system will slowly oscillate away from the point 0.67, an oscillatory repellor (figure 4.27A). So the value of $r = 1.0$ is a kind of break point for the parameter r. The system behaves qualitatively differently depending on whether r is greater than or less than 1.0.

When $r = 1.0$ we refer to the model as structurally unstable, meaning that the slightest change in the parameter will yield a qualitatively distinct form of behavior of the system in general. It can be generally said that structurally unstable models have no place in ecological systems (May 1974). Yet, there is a practical point about structural instability that should be appreciated. If, for example, the value of r in the above example is 1.00001, the model is structurally stable, according to the mathematical definition. But for practical ecological purposes, we might expect evolutionary changes or environmental changes to induce secular changes in r on the order of 0.01 (say) during a meaningful period of time. This means that from a practical point of view, $r = 1.00001$ is practically the same thing as $r = 1.0$. A system that is "just barely stable" may, through relatively small changes in parameters, change its qualitative behavior. Because of this qualitative sensitivity to small parameter changes, the system may be referred to as practically structurally unstable. In this sense the notion of structural instability may be a very important concept for population ecology. The pure mathematical notion of structural instability (an infinitesimally small change in

parameter produces a qualitative change in behavior) may make structurally unstable models unsuitable for ecological phenomena, but practical structural instability may be extremely important.

Whatever their biological significance, there can be no doubt that
points of structural instability play a crucial part in analyzing the overall qualitative behavior of models. Such points are referred to as bifurcation points. Returning to the logistic equation (equation 5), consider
the case of $r = 2.0$. In figure 4.28 three situations are illustrated: $r < 2.0$,
$r = 2.0$ and $r > 2.0$. In the same sense as above, $r = 2.0$ appears to be a
structurally unstable situation. The smallest reduction from the value
of 2.0 yields a population that asymptotically approaches a point attractor, while the smallest increase from the value of 2.0 yields a population that oscillates toward a point attractor. Thus, the model is structurally unstable when $r = 2.0$. Another structurally unstable point is
illustrated in figure 4.29. In this case, the middle figure (figure 4.29B) is
a graph of the logistic with $r = 3.0$. If r is decreased slightly, the figure in
figure 4.29A emerges, and the behavior of the system is damped oscil-

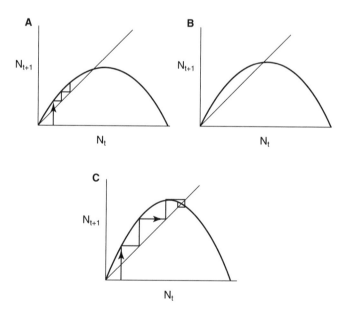

Figure 4.28. Graphs of equation 5 (the logistic equation), illustrating the
structurally unstable configuration obtained when $r = 2.0$. (A) $r < 2.0$, leading
to a stable node (nonoscillatory point attractor). (B) $r = 2.0$, the bifurcation
point. (C) $r > 2.0$, leading to a stable focus (oscillatory point attractor).

$$N_{t+1} = rN_t(1 - N_t)$$

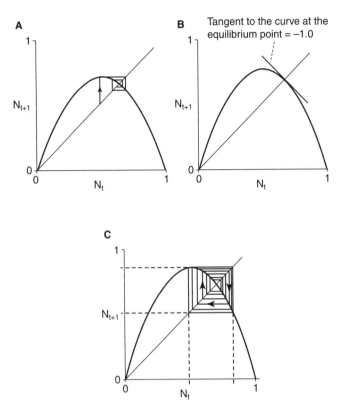

Figure 4.29. Graphs of equation 5 (the logistic equation), illustrating the structurally unstable configuration obtained when $r = 3.0$. (A) $2.0 < r < 3.0$ (see figure 4.13C). (B) $r = 3.0$. (C) $r > 3.0$, leading to a two-point periodic attractor.

lations to a point attractor. If r is increased from 3.0, the figure in figure 4.29C emerges, and the behavior of the system is permanent oscillations with a period of 2, which is to say the population, no matter where it is initiated, eventually oscillates forever between two fixed values, indicated with dashed lines in the figure. Again, the model with $r = 3.0$ is structurally unstable in the sense that a qualitative change in the behavior of the system emerges when the parameter is changed ever so slightly. Furthermore, $r = 3.0$ is also known as a bifurcation point. The system went from a single stable point attractor (oscillatory) to a stable period two cycle. This type of bifurcation is known as a period doubling bifurcation (in the context of continuous systems:

i.e., with differential equations, the same qualitative arrangement is known as a Hopf bifurcation).

A very different type of bifurcation may arise in more complicated models. Consider, for example, the case modeled above of a constant population of predators in the system (equations 6a and 6b). With the appropriate choice of parameters, the situation in figure 4.30 may arise. Once again the center graph (figure 4.30B) is a bifurcation point, in that one small change in a parameter may create the situation in figure 4.30A, while a change in the other direction might result in the situation illustrated in figure 4.30C. Note that two new equilibrium points have been created (or destroyed) in this bifurcation, one an attractor (referred to sometimes as a node) and one a repellor (referred to sometimes as a saddle). This type of bifurcation is variously referred to as a

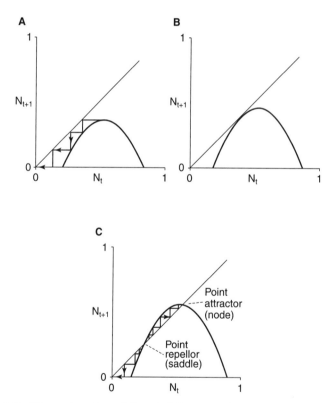

Figure 4.30. Illustration of a saddlenode bifurcation. (A) A single attractor at 0. (B) The bifurcation point. (C) After the bifurcation, there are a repellor (saddle) and an attractor (node), indicating that the bifurcation was of the saddlenode type.

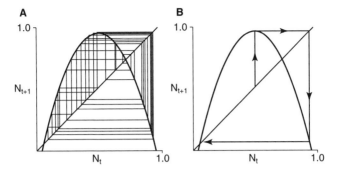

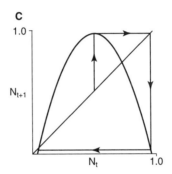

Figure 4.31. Illustration of a basin boundary collision. (A) Alternative attractors, one at 0, the other a strange attractor. (B) Bifurcation point. (C) After the basin boundary collides with the strange attractor, the trajectories that had been part of the attractor are now just part of the basin of attraction for the point attractor at 0.

saddle/node bifurcation or a blue sky bifurcation (since two equilibrium points appear "out of the blue").

The above two bifurcation types (period doubling and saddle/node) are both characteristic of changes in point attractors and repellors. In the case of the period doubling bifurcation, the bifurcation itself shifts the model from a point attractor to an oscillatory attractor, but before the bifurcation, the equilibrium is a point attractor. Other types of bifurcation may involve strange attractors and tend to be far more complicated. Consider, for example the constant predator model (equations 6a and 6b). In figure 4.31 this model is plotted first (figure 4.31A) in such a way that there are two attractors, to the left a point attractor at zero and to the right a strange attractor, and the two are separated by a separatrix. A slow change in parameter results in the condition illustrated in figure 4.31B. A further small change in parameter causes the strange attractor to collide with the basin of the point attractor, eliminating the strange attractor entirely (figure 4.31C). What used to be parts of the trajectory of the strange attractor are now simply trajectories within the basin of attraction of the point attractor at zero. This sort of bifurcation phenomenon is known as a basin boundary collision because the basin of one attractor (the lower point attractor at zero) collides with the dynamic boundary of the strange attractor.

These examples, two structural instabilities associated with point attractors (period doubling and saddle/node bifurcations) and a structural instability associated with a strange attractor (basin boundary collision), are but a few of the possibilities. Other structural instabilities are possible with more complicated models but are beyond the scope of this text.

Structural stability, then, is a very different form of stability from those discussed earlier, which involved points or cycles. It involves the structure of the model as a whole and its qualitative behavior.

Bifurcation Diagrams

Nonlinear models are frequently resistant to traditional analytical treatment. The one-dimensional maps that have served to illustrate basic model behavioral forms in this chapter are really only heuristic devices, enabling a partial understanding of the kinds of underlying structures that may give rise to complicated behaviors in more realistic models. For example, in more complicated predator–prey models, it is sometimes possible to construct an approximate one-dimensional map that captures the qualitative behavior of the more complicated model (e.g., Schaffer 1985). The basic structure of the one-dimensional model is then much easier to comprehend than the complicated model (a model of a model, so to speak).

One way of examining the general behavior of a complicated model, when traditional analytical procedures are unavailable (i.e., because of the complexity of the model, it is not possible to treat them analytically), is the bifurcation diagram. The various forms of bifurcation already described in this chapter (i.e., points of structural instability) are sometimes easily visualized in a "bifurcation diagram." Consider the standard logistic map:

$$N_t = rN_t(1 - N_t)$$

As shown earlier, this model will have either a periodic or a strange attractor whenever $3 < r < 4$. Which is to say, 3 is a bifurcation point. That means that when $r < 3$, the solution is a simple equilibrium point, and when $r > 3$ the solution is periodic. When r is only slightly greater than 3, the periodic solution is a "period two" attractor, which is to say the system oscillates between two points, exactly the same two points, forever. Consider, for example the value of $r = 3.2$. If we begin with $N_t = 0.8$, we can easily calculate,

$$N_2 = 3.2\ N_1(1 - N_1) = 3.2\ (0.8)(1 - 0.8) = 3.2\ (0.8)(0.2) = 0.512$$

Then we can substitute 0.512 to compute N_3, as,

$$N_3 = 3.2\ (0.512)\ (1 - 0.512) = 3.2\ (0.512)\ (0.488) = 0.8$$

which is the number we originally started with. Thus we see that in this case the model will oscillate between 0.512 and 0.8 in perpetuity. This is a period two (two values that are repeatedly visited) attractor. Note that this system is not neutral; we made a point of starting it exactly on one of the points of the two-point cycle.

Thus we had, when r was slightly less than 3, a single equilibrium point, whereas when r became slightly greater than 3, we had a period two attractor. Without further proof (the interested reader can verify any of this with some simple experiments on a hand calculator), if r becomes yet larger, the period two attractor converts into a period 4 attractor and if it becomes even larger the period 4 attractor converts into a period 8 attractor. In figure 4.32 this process is illustrated by way of a graph of N^* against r, where N^* is either the equilibrium point or one of the repeated points of the attractor.

In figure 4.32 the points referred to above are plotted, but they also are connected with dashed lines. It is intuitively obvious that the dashed lines represent an approximation to what the intermediate values of N^* would be, and it is clear that the dashed lines bifurcate at critical points. The diagram in figure 4.32 is thus referred to as a bifurca-

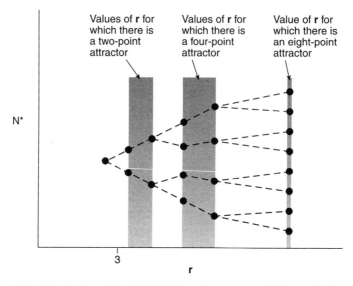

Figure 4.32. Illustration of the basic bifurcation process. Values of r that result in various attractor types are illustrated. The points are equilibrium points or, rather, equilibrium cycles. Each point represents a point on the y axis that will be part of the attractor at the particular value of r. For example, there are two points at $r=3$, which, as discussed in the text, are the two points of the two-point cycle. Dashed lines connect these points in what is likely to be the intermediate values giving rise to the different attractors. Note the bifurcating nature of the picture, giving rise to the appellation bifurcation diagram.

tion diagram, and it is an important tool that is commonly used to study complex models.

Rather than choosing particular values of r and calculating the values of N^*, we can simply calculate N^* for all of the values of r, incrementing by some small amount. If we do this process for the logistic map, we obtain the graph presented in figure 4.33.

The various attractors described in figure 4.32 are clearly visible. Also visible is the bifurcation event at $r=4$, fully explained in an earlier section. By examining such a bifurcation diagram, one frequently can gain an overall picture of how the model behaves. In this case, there is a clear cascade of period doubling events, from one to two to four to eight. In figure 4.34 a part of the bifurcation diagram is expanded (that is, the diagram is made for $r=3.5-3.7$.

Note that the period doubling cascade is now visible for periods four, eight, and sixteen. What happens is what one would expect, for

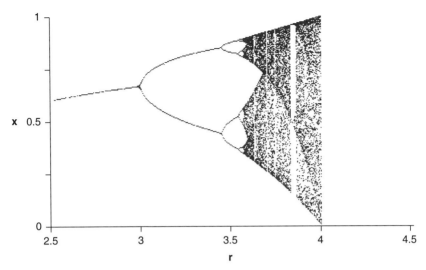

Figure 4.33. Bifurcation diagram for the logistic map. Note the similarity to figure 4.32. There is a clear period doubling bifurcation at $r = 3$. At $r = 3.5$ another period doubling has already occurred, and the system is in a four-point attractor. At r slightly larger than 3.5 another doubling occurs and the period eight attractor is visible. Note also the period three "window." At $r = 4$ there is another bifurcation, as described earlier in the text.

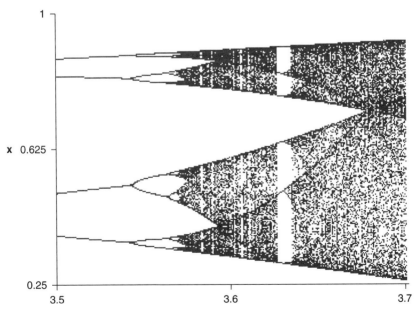

Figure 4.34. Close-up of part of the bifurcation diagram of figure 4.33.

the most part. The periods keep doubling, and the change necessary in r to get to the next doubling keeps decreasing. Eventually, there has been such a massive period doubling that a remarkable point is reached in which one can simultaneously get all possible periods as well as an uncountable number of aperiodic (i.e., never settling down to permanent values) attractors. This is the point at which the system is usually referred to as chaotic. The manner in which chaos is approached here is referred to as the period doubling route to chaos (other, qualitatively distinct, forms also exist but are beyond the scope of this text). Just how complicated this behavior is can be appreciated by noting that there is a "window" in the original diagram (figure 4.33) in which there is an attractor of period 3. Where does a period three attractor come from if the sequence goes 1, 2, 4, 8, 16, . . .? Similarly, in figure 4.34 there is a window with a clear period six attractor. Where does that come from? Suffice it to say here that the explanation is mathematically complicated and beyond the intentions of this text. But biologically all it means is that the system is extremely unpredictable in the range of about $3.58 < r < 4$. This extreme unpredictability is what suggested the term *chaos*.

As an example of the utility of bifurcation diagrams, consider the *Tribolium* model presented in chapter 3. Recall the population was divided into larvae, pupae, and adults, and the nonlinear projection matrix model was given as,

$$
\begin{vmatrix} L_{t+1} \\ P_{t+1} \\ A_{t+1} \end{vmatrix} = \begin{vmatrix} 0 & 0 & f_1(L_t, A_t) \\ p_{lp} & 0 & 0 \\ 0 & f_2(A_t) & p_{aa} \end{vmatrix} \begin{vmatrix} L_t \\ P_t \\ A_t \end{vmatrix}
$$

where L is number of larvae, P is number of pupae, and A is number of adults. The functions f_1 and f_2 stipulate the nonlinear effect of cannibalism on the population. The functions are given as:

$$
f_1 = \frac{b}{e^{c_1 L_t + c_2 A_t}}
$$

and

$$f_2 = \frac{1}{e^{c_3 A_t}}$$

which are intended to incorporate the biological fact of cannibalism. This model is in fact quite daunting to solve analytically, and it does not lend itself to any obvious intuitive or heuristic explanation. But if one generates a bifurcation diagram, as Costantino and coworkers (1997) did, the diagram as pictured in figure 4.35 is obtained. This bifurcation diagram was obtained by performing a series of experiments to estimate the parameters of the model and then substituting those values into the model, fixing all parameters except c_3. The parameter c_3 represents the per adult survival of larvae and is a parameter that Costantino and colleagues could experimentally manipulate in the laboratory. They next chose particular values of c_3 that represented various dynamic situations, as indicated by the arrows on the top of the bifurcation diagram, and set up laboratory cultures corresponding to those particular values of c_3. Their results are shown in figure 4.36. The open circles are the experimental results, and the closed circles are the expected results based on the model. The six graphs correspond to the position of the arrows in figure 4.35. There is a remarkable correspondence between expected and observed. What is even more remarkable

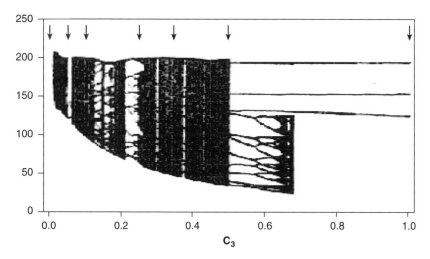

Figure 4.35. Bifurcation diagram of the *Tribolium* model. The bifurcation parameter is c_3, and the variable plotted is the total population size. Reprinted with permission from Costantino et al. (1997). © 1997 American Association for the Advancement of Science.

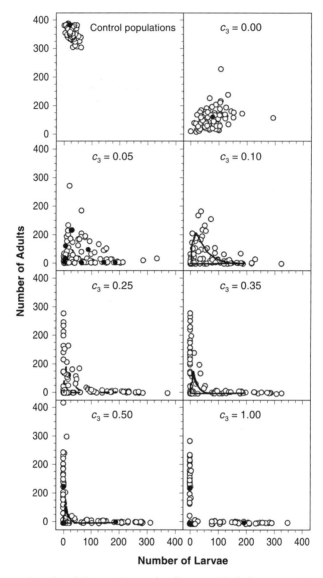

Figure 4.36. Results of Costantino and colleagues' *Tribolium* experiment (1997). Upper left-hand graph is the control; all others correspond to the values of c_3, indicated by the arrows in the upper part of figure 4.35. Open circles are the observed values; closed circles, the expectation based on the model. Reprinted with permission from Costantino et al. (1997). © 1997 American Association for the Advancement of Science.

is that the biologists in this team of investigators first thought that several of the theoretical outcomes would be impossible to achieve in the laboratory. But the data speak for themselves. Here, the bifurcation diagram was critical to the evaluation of the model to the point that predictions from the model came directly from the bifurcation diagram. The more traditional technique of comparing changes in the variables over time with the predictions over time generated by the model and by experiments could never have provided such a strong and elegant test of the model as the analysis of the bifurcation diagram did.

Concluding Remarks

In this chapter, we have dealt with a panoply of subjects related to the analysis of population models. These sorts of methods are currently the subject of intense investigation, and certainly this chapter will seem outdated within a few years. Nevertheless, the concepts are important and are playing an increasingly important role in population modeling these days. A minimal introduction, such as was provided in this chapter, was thus thought to be essential.

We might cite one practical conclusion. Equilibrium and stability, long the bastions of creative thought in the search for a theory of ecosystems, may need to be abandoned, at least in terms of their classical meaning. For example, a dynamic system in chaotic behavior, such as the illustration in figure 4.24B or 4.31A, is, by classical standards, unstable, and in a strict mathematical sense, unpredictable. Nevertheless, it has clear boundaries to its behavior and in another sense is quite stable—within the dynamic boundaries of its own region. Which sense is important as far as an ecosystem manager is concerned? Which sense is important in terms of understanding the ecosystem? Which sense is important in the context of natural selection?

Such analysis may suggest resolutions for various paradoxes in ecology. For example, the conundrum involving diversity—i.e., diversity is thought to generate stability, yet some highly diverse systems are thought to be quite fragile—can easily be resolved. Perhaps the stability thought to result from diversity actually refers to regional stability with a broad basin of attraction. The fragility might refer to the possibility that the basin itself could become smaller as diversity is reduced, increasing the likelihood that the basin could be traversed and the integrity of the system thus breached. Whether this is actually true of highly diverse systems is not the point here. Rather, these concepts suggest the way that some, perhaps many, natural historians have thought about this issue when they ponder the relationship between diversity and stability. Perhaps it is really what the romanticists mean when they construct poetry about biodiversity and its importance in the world.

In the sort of truly complex ecosystems likely to be encountered in the real world, the examples in this chapter will seem simplistic. Yet the Newtonian notion of point stability remains an essential framework for many thinkers in the field of ecosystem dynamics, if only tacitly so. The simple one- and two-dimensional examples of this chapter are intended to introduce the notion of regional stability and the various complexities associated with it. Any real-world system will be multidimensional and ultimately must be represented in hyperspace. Figure 4.37 presents a "collapsed" hyperspace, a fictional two-dimensional representation of a multidimensional space. Figure 4.37A shows a system with five unstable points, yet there are two attractors, which contain two of the five unstable points. The attractors are regionally stable. By classic definitions of stability, this would be a very unstable system indeed. Traditional analysis of neighborhood stability would determine that there were four equilibrium points, all of which would be unstable. Yet almost anyone would agree that this system is more "stable," in some vague ecological sense, than the alternative system illustrated in figure 4.37B. This system has two stable points (one at zero), yet intuitively most would regard it as less stable than the one in figure 4.37A. The notion of regional stability more clearly encompasses what most workers in ecology would describe as stable, and a regional attractor is similarly closer to intuitive notions of equilibrium than the single points of the neighborhood stability sense.

The notion of structural stability represents a totally different idea of stability than either the neighborhood or the regional sense, and in many ways it is probably similar to what environmentalists really mean when they refer to a "stable" system—a system that shows particular characteristics and will continue to show those characteristics even if small changes in conditions occur in the environment. So, for example, when the local environment changes such that a crop pest develops a locally elevated population density, the natural enemies of the traditional agroecosystem may respond by exerting a control over that

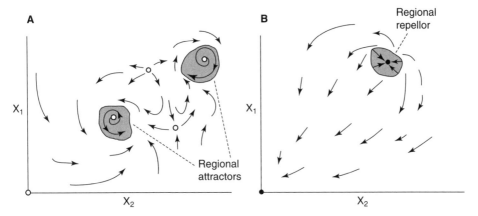

○ Point repellor
● Point attractor

Figure 4.37. Theoretical situations. (A) Five point repellors and two regional (strange) attractors coexist. (B) Two point attractors and one regional repellor coexist. The point attractors are illustrated by closed circles and the point repellors by open circles.

temporarily elevated population. A parameter change had occurred (the local environment that resulted in the elevated population density of the pest), but the system was structurally stable. No large change in its behavior resulted from this change in parameter.

Yet, points of structural instability, especially bifurcation points, are sometimes exactly what we are looking for when we aim to understand or design ecosystems. Could it be, for example, that the conventional production techniques for producing processing tomatoes in California simply represent a syndrome of production, that another syndrome (the organic method, for example) might emerge if the parameters are changed somewhat, and that strategists aiming to convert to organic production might very well look for that break point, the bifurcation point that will carry the entire system into the organic mode? On the other hand, gradual changes in parameters may lead to a basin boundary collision in which a system originally held within a bounded attractor that represents a desired state of the ecosystem is engulfed into the basin of an attractor that includes undesired states. This phenomenon has been suggested as a possible mechanism for sudden extinction in natural populations (McCann and Yodzis 1994). This somewhat philosophical point will be left for the reader to ponder. Suffice it to say that the notion of structural stability is crucial in many ways to the understanding and designing of ecosystems.

Throughout this chapter we have minimized use of the word *chaos*, even though much of what is included is closely related to the field commonly known as chaos theory. We have done this because the word *chaos* has been something of a misnomer, leading to some confusion about the implications of chaotic behavior. The chief source of confusion comes from what has perhaps been an overemphasis on one particular aspect of chaos, sensitive dependence on initial conditions, especially in the popular literature on the subject. This particular characteristic is actually not even uniquely characteristic of chaos; unstable points exhibit the same phenomenon. However, the persistence of systems even though chaotic, coupled with the property of sensitive dependence on initial conditions, leads to the intuitive notion that they are inherently unpredictable. Although this notion is true in a narrow technical sense, it is certainly not the most important aspect of strange attractors (a better name than chaotic attractors).

Consider, for example, a tornado (Vandermeer and Yodzis 1999). That is most likely a chaotic object, an example of a strange attractor. It represents sensitive dependence on initial conditions in the following sense: Consider two particles of dust within the tornado. If they are very close to one another at one point in time, that fact has no bearing on where they will be with respect to one another in the near future. And how close they are now is not at all correlated with how close they will be in the future. So the future location of each dust particle is dependent on exactly where it is now and may change very dramatically with only a very slight change in its position now. That is sensitive dependence on initial conditions. But in fact that is not the interesting thing about a tornado. It is shaped like a funnel—that is, it has a morphology—and the whole thing travels along the ground wreaking devastation wherever it goes; that is, it has a behavior. Furthermore, if you see one coming toward you, it is really quite a good idea to get out of the way, even though it is chaotic and therefore "unpredictable." Sensitive dependence on initial conditions, the key idea of chaos's unpredictability, refers only to the behavior of those dust particles inside the tornado. What is truly interesting about a tornado, what we wish to know and even predict, is perfectly knowable and predictable—thankfully. As so eloquently stated in a recent summary of complexity theory, "First, control of natural phenomena begins to slip out of the grasp of observers, both because sensitivity to initial conditions severely limits the possibilities for prediction and control and because emergent properties of complex systems are unpredictable from a knowledge of parts, . . . Second, these emergent properties can nevertheless be made intelligible in terms of appropriate descriptions of the processes involved, by using high-level concepts that capture their essential aspects" (Solé and Goodwin 2000).

It is probably not of particular interest to ask whether a system is "formally" chaotic. An extremely complicated periodic attractor is, for all practical purposes, equivalent to a strange attractor anyway. If the behavior of the system is bounded and there is an instability within the bounded region, for all practical purposes it may be treated as if it were a strange attractor. Granted, there are cases of rather simple periodic attractors that can be analyzed in a traditional fashion. But for many of the system behaviors that we can expect of populations embedded in ecosystems, their behavior is likely to be very complicated, more like strange attractors than simple points or limit cycles. This does not mean they cannot be understood, any more than it means that a tornado has no shape. The focus should be not on the unpredictability but rather on the morphology of the attractor: Where are its boundaries? Is it periodic-like? Does it have dense and less dense regions? What is its overall shape? and so forth—the "appropriate descriptions" referred to in the above quotation. Similar to the morphology of organisms, there is no one defining feature of the morphology of strange attractors.

In studying the morphology of attractors, we have here presented what seem to be some key principles. Where are the boundaries of a regional attractor? Is it structurally stable (in a practical sense)? What are the natures of nearby bifurcation points? Where are the basins? These and similar questions are likely to be the questions we are able to answer about ecosystems in general, and, furthermore, we suggest they are the ones we are more interested in answering anyway, in pursuit of understanding and designing ecosystems.

5
C H A P T E R

Patterns in Space and
Metapopulations

A bacterial population increases exponentially, at least for a short period of time, on a nutrient agar substrate. A population of rodents is maintained in a region but with dramatic shifts in numbers from year to year, seemingly like the chaotic patterns reflected in some of the simple models already discussed. These are patterns in time. Acacia trees in the African savannas tend to occur in thickets, where hundreds of individuals are concentrated in relatively small areas, and between these concentrations it is uncommon to find even a few individuals (usually the concentrated areas are near areas where water is close to the surface). This too is a pattern, but in space rather than in time. In recent years, the analysis of this sort of pattern has become increasingly important in population ecology, and extremely sophisticated tools have evolved to aid in this analysis (see, for example, Durrett and Levin 1994, 1998). Here we deal with only the most elementary aspects of spatial analysis. The more complicated analyses are indeed quite complicated and beyond the intended scope of this book.

Rather than expand on what are essentially measurement methodologies (the more complicated analysis mentioned above), we include in this chapter an introduction to a way of thinking about space that has become a whole field of ecological research unto itself: the subject of metapopulations. It may seem unusual that this subject is included in a chapter on patterns in space, but in fact the initial forays into metapopulation dynamics stems from observing patterns similar to the one described about the acacia trees. When populations are highly clumped, with dense patches in a matrix of rarity, one begins thinking

Epiphytic bromeliads create habitat patches within which a host of organisms live. Such a patchy distribution of organisms is frequently an example of a metapopulation.

at two levels: (1) the analysis of the population dynamics of one of the patches, which is essentially what has been presented in previous chapters; and (2) the analysis of what happens between patches, considering the collection of patches as the population. This latter focus is the subject of metapopulations.

We begin with some simple spatial concepts. Consider the illustration in figure 5.1. Figure 5.1A is the distribution of white oak trees in a 100×70 m plot at the E. S. George reserve. There seems to be little pattern to the data: it is almost like the pattern you might get from a shotgun blast. In figure 5.1B is the distribution of witch hazel trees in the same plot, illustrating an obviously different pattern. The witch hazels are clumped in a couple of parts of the plot, whereas the white oaks are dispersed around the plot. How should such differences be quantified? Are there different spatial levels at which quantification should be focused? What are the dynamic consequences of the difference? We begin with the simple problem of how to measure such spatial patterns.

Consider the pattern of birds on a telephone line. At one extreme are

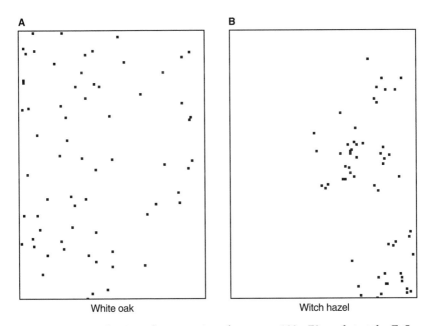

White oak   Witch hazel

Figure 5.1. Distribution of two species of tree on a 100×70 m plot at the E. S. George Reserve near Ann Arbor, Michigan.

Figure 5.2. Birds on a telephone line. (A) Clumped. (B) Evenly dispersed.

the distributions of parakeets, who love being together and tend to clump up on the line (see figure 5.2A). At the other extreme are black-birds, who continually fight with one another and wind up evenly distributed on the line (see figure 5.2B). What sort of an index will reflect the obvious difference between these two patterns?

The Poisson Distribution

The usual way of approaching this problem is by assuming a sort of null model, at least initially. What if the birds land on the line at random and we measure their positions before they either clump up (as do the parakeets) or disperse themselves out (as do the blackbirds). Divide the telephone line into very small units, so small that only a single bird can fit in a unit. There is some probability that a given unit will be occupied (mainly determined by the population density and behavior of the birds in the vicinity). That probability is p. We now ask the question, What is the probability that exactly r units will be occupied on a length of telephone line that contains n spaces (remember, at this point each space can contain only a single bird). For example, suppose there are only three spaces and $p = 0.4$, what is the probability that exactly two of the three spaces will be occupied? All possibilities are shown in figure 5.3. In other words, the probability that two of the three spaces on the telephone line will be occupied is $(3)(0.4)^2(1 - 0.4)$. We can generalize this idea by noting that the probability that r of n spaces will be occupied if p is the probability of such space being occupied is

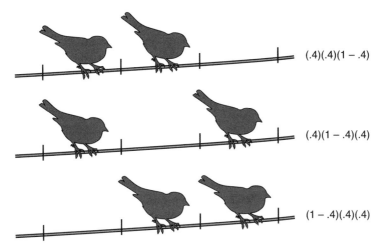

$(.4)(.4)(1 - .4)$

$(.4)(1 - .4)(.4)$

$(1 - .4)(.4)(.4)$

Figure 5.3. All possible ways that two birds can fit on three places on a telephone line.

$$Cp^r(1-p)^{n-r}$$

Where C is a constant equal to the number of ways r things can fit into the n locations. In general, the constant C is given as

$$\frac{n!}{(n-r)!\,r!}$$

In the preceding example, $r = 2$ and $n = 3$, which gives

$$\frac{n!}{(n-r)!\,r!} = \frac{(3)(2)(1)}{[(1)][(2)(1)]} = \frac{6}{2} = 3$$

Thus, in general, the probability that exactly r locations will be occupied out of a possible n for which the probability of being occupied is p is,

$$P_r = \frac{n!}{(n-r)!\,r!}\,p^r\,(1-p)^{n-r} \tag{1}$$

Which is the bionomial probability: the rth term in the expansion of $(1-p)^n$.

Figure 5.4. Dividing the telephone line into segments (sections), each of which could, theoretically, contain a large number of birds.

Now suppose we have a very long telephone line and we divide it up into relatively large sections (so that some large number of birds could fit into each section), something like figure 5.4. Each of these sections is made up of a large number of small segments, each of which can contain a single bird. The probability that exactly r of the subsections in a given section will be occupied by a bird is the same as the probability that the section will contain r birds, and we already know that that is given by the binomial. However, now we are assuming that there are a very large number of subsections in a given section of the telephone line. Effectively, this means that n is very large compared with r. And this is the trick. Assuming n is very large with respect to r enables us to derive an exact formula to describe a random distribution (assuming that each of the r subsections is occupied at random for all of the segments is effectively the definition of a random distribution of birds on the telephone line). From equation 1 we have,

$$P_r = \frac{n!}{(n-r)!\,r!}\,p^r\,(1-p)^{n-r} = \frac{n(n-1)(n-2)\ldots(n-r+1)}{r!}\,p^r\,(1-p)^{n-r}$$

and, allowing n to be very large with respect to r, we can see that $n-r$ (the exponent in the last term) is approximately equal to n, and $n(n-1)(n-2)\ldots(n-r+1)$ is essentially the same as n multiplied by itself r times (since n is so large in comparison with r). Thus we have,

$$\frac{n^r\,p^r\,(1-p)^n}{r!}$$

Symbolizing np with μ, we obtain,

$$\frac{\mu^r \left(1 - \dfrac{\mu}{n}\right)^n}{r!}$$

Recall from basic calculus that

$$\lim_{n \to \bullet} (1 - x/n)^n = e^{-x}$$

Thus, if we allow n to become very large,

$$P^r = \frac{\mu^r e^{-\mu}}{r!} \tag{2}$$

which is the equation for the Poisson series, where μ is the mean number of birds per segment (or, more generally, the mean number of elements per quadrat).

Applying this formula to the original data on trees at the E. S. George Reserve, we redraw the plots with grid lines, as in figure 5.5. Now count the number of quadrats that have 0 trees in them (49 in the

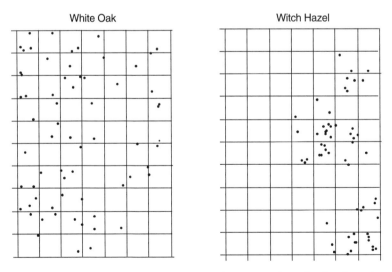

Figure 5.5. Same graph as figure 5.1 but with 10-m grid marks dividing the plot into 70 10×10 m quadrats.

case of witch hazel), then the number of quadrats that have 1 tree in them (8 in the case of witch hazel), then the number that have 2 trees, and so forth. The total number of trees in all the quadrats divided by the number of quadrats (in this case 70) gives the value of μ (the mean number of trees per quadrat). With this estimate of the mean, we apply equation 2 and multiply each probability by 70 (the total number of quadrats) to get the expected number of quadrats for each category of tree density. We then compare the expected with the observed, as is done graphically in figure 5.6. As can be seen, the observed values for witch hazel are significantly different than the expected, but the plot for white oak is very much as expected; it more or less follows what would be expected for a random distribution. There is a slight tendency for hyperdispersion in oak since there are fewer zero quadrats

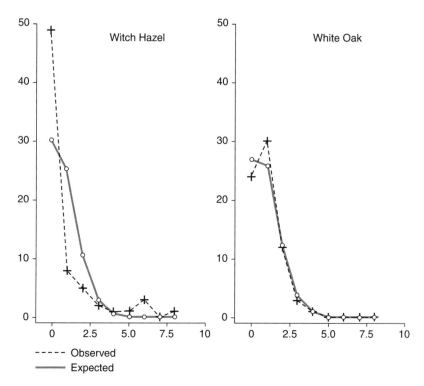

Figure 5.6. Comparison of expected (based on Poisson expectation) with observed density for the witch hazel and white oak trees in the plot at the E. S. George Reserve.

and more singletons than would be expected. Testing the hypothesis of randomness is thus simply a matter of using a chi-square or similar statistic to compare expected with observed.

The Poisson distribution has an extremely useful property. In a perfectly random distribution, the mean and the variance of the number of individuals per quadrat are identical. Consequently, one can use the ratio of variance to mean as a measure of "clumpiness." So, for example, taking the above examples of trees in the E. S. George Reserve, for white oak the mean number of individuals per quadrat is 0.954. The number of quadrats with zero trees is 24; the number with one tree is 30; the number with 2 is 12; the number with 3 is 3; and the number with 4 is 1. There are no quadrats with 5 or more individuals. So the variance will be $24(0.957-0)^2 + 30(0.957-1)^2 + 12(0.957-2)^2 + 3(0.957-3)^2 + (0.957-4)^2 = 56.872$, divided by the number of quadrats less 1 (69), which gives 0.824. The variance-to-mean ratio then is 0.824:0.954 = 0.863, which is very close to the theoretical expectation of 1.00, which is what we would expect from an examination of the graph in figure 5.6. Making the same calculation for witch hazel, we find a variance of 2.763 in comparison with a mean of 0.843, making the variance-to-mean ratio equal to 3.27. Thus, we conclude that white oak is random, perhaps a bit more evenly dispersed, whereas witch hazel is highly clumped.

The variance-to-mean ratio is frequently taken as a quantitative measure of degree of clumpiness. A ratio that is greater than 1 indicates a clumped distribution; less than 1 indicates an evenly dispersed distribution; and equal to 1, a random distribution.

The Question of Scale

It is the bane of the empiricist to discover that the degree to which a population is clumped in space depends on the scale that is used for analysis. Consider a fictitious population of giant hornwarts censused on a 10×14 m grid, illustrated in figure 5.7A. The plants are obviously clumped, but the definition of a clump may very well depend on how you look at the population. Are there just the obvious small clumps, or are the clumps themselves clumped in some way? Applying the simple formulae introduced above to this population gives the expected results. Constructing quadrats as in figure 5.7B, we compute the mean as

A

B

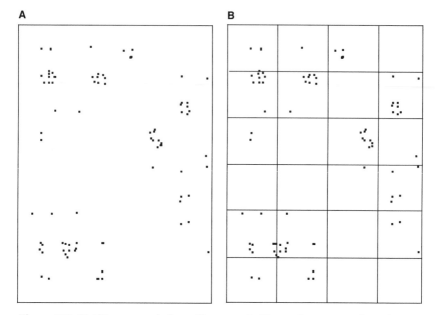

Figure 5.7. Fictitious population of hornwarts illustrating strong clumping.

3.21 and the variance as 13.216, giving a variance-to-mean ratio of 4.117, indicating strong clumping in this population, as expected. However, if we change the size of the quadrat used for analysis, as in figure 5.8, the results also change. Consider the quadrats in figure 5.8A. Using these small quadrats, we compute a mean of 0.278, a variance of 0.503, and a variance-to-mean ratio of 1.80. Thus, with this new quadrat size, our estimate of clumping is more than twice what it was with the larger quadrat size. Worse, consider the quadrats in figure 5.8B. Here, the mean is 1.114, the variance is 1.088, and the variance-to-mean ratio is 0.976, which means that the population is actually evenly dispersed (though not really significantly different from random). Thus, depending on the choice of quadrat, the same population can be extremely clumped, mildly clumped, or random! This is the problem of scale when dealing with spatial patterns.

If we compute the variance-to-mean ratios for a variety of quadrat sizes and graph the ratio against the size (one side) of the quadrat, we obtain, for this example, the result illustrated in figure 5.9. It might be obvious to conclude that the population is clumped, but it is not clear what further conclusions could be drawn other than that the size of the

A   B

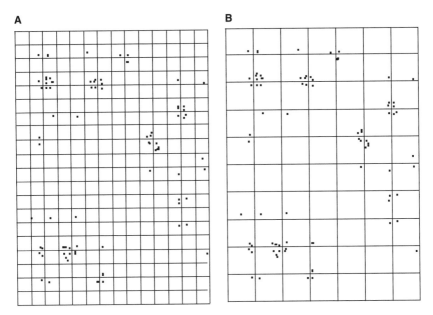

Figure 5.8. Two quadrat sizes superimposed on the population of hornwarts of figure 5.7.

sampling unit is critical in determining whether the spatial pattern is more or less clumped. Diagrams of this sort are frequently used to try to sort out further some of the spatial structure in the population. For example, in figure 5.9, we see a dramatic increase in the variance-to-mean ratio at about 1.5 m, another increase between about 2.25 and 3.5 m, and a continual increase after about 4.5 m. It is tempting to interpret such a pattern as suggesting three levels of clumping. And, with a little imagination, this pattern of hornwarts can be forced into such an interpretation. In figure 5.10A, for example, we illustrate two sizes of clumps: those large ones corresponding to the right end of the graph in figure 5.9 and the small ones corresponding to the left end of the graph. In figure 5.10B we attempt to find the intermediate clumps suggested by the variance-to-mean ratio analysis.

The question of scale is only one of the very complicated problems one must face when dealing with spatial pattern. Even the problem of describing the clumpiness of populations is complicated, as illustrated above. Add the further complication of developing dynamic models (i.e., models in which the population changes through time) that are

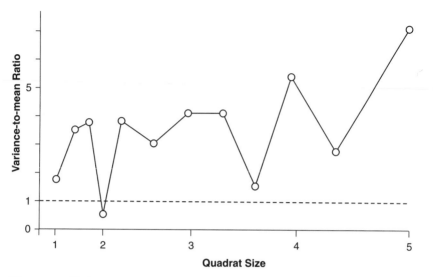

Figure 5.9. Variance-to-mean ratio plotted against the size of the quadrat for the data of figure 5.7.

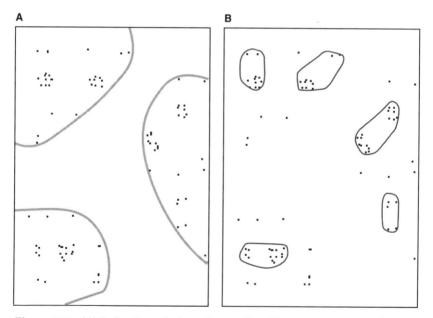

Figure 5.10. (A) Indication of where the small and large clumps are in the hornwart example. (B) Indication of where the intermediate clumps might be in the same example.

explicit spatially, and we face a gauntlet of technical difficulties. A great deal of work has been done on this problem, but it is far beyond the intended scope of this book. The reader who intends to seriously tackle problems involving spatial distributions should not think that the presentation here is adequate to his or her needs. Indeed, as we stated initially, this presentation is intended as an introduction to only the most basic ideas.

Metapopulations

The construction of classic population models (e.g., chapters 1 and 3) tacitly assumes a relatively uniform distribution of individuals in the population. We assume that most individuals in the population are capable of interacting with and, in the case of a sexual population, mating with all other individuals in the population (what biologists call a panmictic population). Yet, even the most casual observer recognizes that most populations in nature are patchy. Habitats tend to occur in a patchwork, and consequently the occupants of those habitats tend likewise to be patchy. This patchiness is obvious in the case of islands, but it is also worth noting that most non-island populations have an islandlike structure when looked at spatially. Recently, there has been considerable interest in the dynamics of such spatially distributed populations because understanding how they work appears to have a variety of practical applications.

Harrison (1991) has provided a useful classification of different forms of spatially distributed populations, reproduced here as figure 5.11. First there is the island type (also called source-sink): the sole source of colonists is a large "mainland," or source area, in which the population is persistent, and there are "islands," or sink areas, in each one of which the population would go extinct if cut off from the mainland source of colonists (figure 5.11A). The next case in figure 5.11 is the panmictic population with a clumped spatial distribution (figure 5.11B). The next is the nonequilibrium "metapopulation" (figure 5.11C), in which there is too little migration to maintain the overall population so it is on its way to extinction (sometimes just referred to as *fragmented*, although this term is also used for a habitat that is literally fragmented, regardless of the behavior of the populations contained therein). In figure 5.11D is a garden-variety source-sink situation (i.e., the same as figure 5.11A), in which the source population is clumped in space.

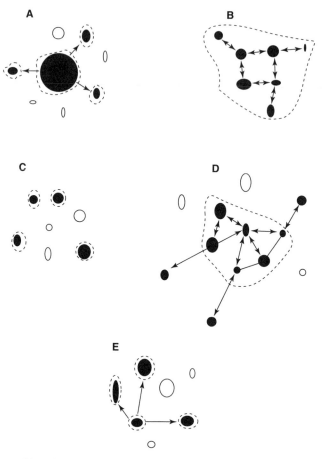

Figure 5.11. Classification of spatially structured populations (according to Harrison 1991). Circles and ovals represent habitat patches; filled circles and ovals are occupied, unfilled are vacant. Dashed lines indicate the boundaries of "populations." Arrows indicate migration and colonization (Harrison and Taylor 1997).

The remaining case is the classic notion of a metapopulation (figure 5.11E), a term first coined by Levins (1969). Much like island biogeography, a metapopulation has subpopulations on isolated islands. But rather than experiencing extinction and recolonization from a mainland or source area, the already occupied islands serve as a generalized source for new migrants. Thus, we expect extinctions to occur on the "islands," but they are inevitably followed by recolonization from one of the islands that contains an extant population. This general concept has spawned an enormous literature in the past two decades. It has become a springboard for a new way of looking at population dynamics, since it breaks completely with the tradition of looking at birth and death rates of populations and begins instead with migration and extinction rates.

The theory of island biogeography (MacArthur and Wilson 1967) was perhaps the starting point for thinking about populations dynamically in the context of spatial structure, although that theory dealt with assemblages of species not specifically with populations. Levins extended the theory of island biogeography. He first conceived of a population distributed in clumps. Key to the idea was that no single clump could form a successful population in perpetuity. If isolated from other clumps, it would become extinct within the area formerly occupied by that clump. The only thing that keeps the overall population (the collection of subpopulations in the clumps) alive is migration from clump to clump. Rather than constructing the model from the typical state variable of population density, Levins conceived of the overall population as a collection of potential clumps, some of which were actually occupied by a subpopulation, others of which were not yet occupied. Then, he speculated that the dynamics of the overall population was one of local populations (subpopulations in the clumps) going extinct and clumps' being recolonized by individuals migrating from other subpopulations.

The basic Levins model of metapopulations begins by analyzing the state variable p, where p is the proportion of patches currently occupied by a subpopulation. Note that the use of p rather than N (population size or biomass) is a dramatic departure from classic population ecology. If p represents the proportion of habitats occupied, the dynamics of the process can be described as a balance of migration and local extinction. Let us begin the analysis by assuming that the extinction rate is zero and p is very small (close to, but not exactly equal to, zero). The rate of increase of p is then likely to be equivalent to an exponential process because the larger p becomes, the more likely it is that more migrants will be available to occupy empty patches. Thus we can write,

$$dp/dt = mp$$

where m is the migration coefficient. This equation represents an exponential process with p as the dynamic variable and predicts that

p should increase exponentially. Since p is a proportion, its maximum value is 1.0, which is analogous to the carrying capacity in the earlier derivation of the logistic equation. As the population approaches its carrying capacity (in this case 1.0), its rate of increase decreases. Indeed, we can expect that the migration rate will really be inversely proportional to the proportion of habitats already occupied, $1-p$. Thus, the above exponential form really should read,

$$dp/dt = mp(1-p)$$

which is, mathematically, identical to the logistic equation. However, in developing this model, we assumed that the extinction rate is zero. If we now add to the model the process of extinction, as if it were an independent death rate in the logistic model, we can write,

$$dp/dt = mp(1-p) - ep$$

where m is migration rate and e is extinction rate. Thus the migration rate term is formulated mathematically as a logistic equation (with carrying capacity equal to 1.0).

This simple model generates important insights. Setting the derivative equal to zero and solving for p, we find the equilibrium state as,

$$p^* = (m-e)/m$$

which immediately suggests a couple of interesting ecological conclusions. First, there will be a positive value of p^* (and thus a viable population) whenever $m > e$. This is not really surprising since it simply says that when migration rate is greater than extinction rate the general population will not disappear. The second conclusion is that as long as the extinction rate is greater than zero, the equilibrium state of the population will always include some empty habitats. As a practical matter for conservation, for example, this conclusion suggests that a strategy for preserving a species that requires the species to be present in all available habitats may not be realistic. If the population normally exists as a metapopulation, it is inevitable that some habitat patches will always be devoid of individuals of the species, and, in fact, preserving empty, suitable patches is an integral part of any successful conservation strategy.

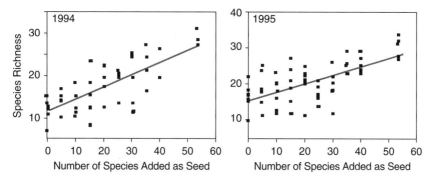

Figure 5.12. Results of Tilman's (1997) seed addition experiments.

Whether a population is truly a metapopulation depends on a variety of criteria. One of the most important is that some potential sites are unoccupied and those unoccupied sites are potentially occupiable. To examine this criteron, a variety of experiments have added propagules to sites that are unoccupied but seem suitable within a region. For example, adding seeds to patches of meadows within a forest matrix, workers in Sweden discovered that indeed many of the species could persist in currently unoccupied sites (Eriksson and Kiviniemi 1999). In another study Tilman (1997) added seeds of up to 54 species to plots in each of 30 prairie openings in native oak savanna. All species occur locally, but not in the experimental plots. A significant increase in species richness with number of species added was found (see figure 5.12). Again, many of the species seem to be limited in distribution by colonization or propagule availability rather than by site suitability.

Assumptions of Metapopulation Models

Six basic assumptions underly the idea of metapopulations.

1. Suitable habitat occurs in discrete patches. This assumption seems obvious in many cases (e.g., meadows surrounded by forest or forests in an agricultural matrix).
2. The population is at least temporarily a reproductive population. For example, in a series of dry meadows in a matrix of forest in Finland, Hanski and Simberloff (1997) found that 60% to 80% of the butterflies spent their whole life in their natal patch,

a finding that provides strong evidence that the patch did not contain just an ephemeral aggregation.

3. Subpopulations have a substantial risk of extinction. In the Finland butterfly example, the largest population had only about 500 individuals (out of 377 populations), and populations with several hundred individuals have been recorded as going extinct.

4. The subpopulations are not too isolated such that recolonization of empty patches is likely. In the Finland butterfly example again, the mean nearest-neighbor distance was 240 m, and the median distance moved by migrating butterflies was 330 m, more than sufficient to allow colonization to balance extinction. On the other hand, it may very well be that many fragmented populations are in a state such that colonization rates will not be high enough to sustain true metapopulation dynamics.

5. The dynamics are not synchronized across subpopulations. This is a major concern in applying metapopulation theory to conservation. If there are correlated extinctions among patches, a uniform dispersal rate may not be able to compensate for the peaks in extinctions.

6. All patches are alike (i.e., equally suitable) both for probability of extinction and colonization and as a source of colonists. Harrison and Taylor (1997) argue that variation among patches is the real case, and it is likely that most metapopulations are closer to the mainland island model (a mainland, by definition, has a zero probability of going extinct and also serves as the sole source of propagules) than to the classic metapopulation model.

These six basic assumptions are the core of current thinking about metapopulations, but other aspects may turn out to be important as research advances. For example, variation in the quality of the matrix in which the patches are located will likely be very important but has seen little attention. Although it is rarely acknowledged explicitly, an additional assumption of metapopulation theory is that the matrix is uniformly inhospitable, an assumption that may not be true. The spatial arrangement of patches is likewise an important issue as is the nature of the boundaries of the patches. These and similar questions are usually considered subjects appropriate for landscape ecology, but as of yet there has been no useful attempt at merging landscape ecology with metapopulation theory (but see Wiens et al. 1993, Vandermeer and Carvajal 2001). Here we examine but two of the complicating issues of metapopulation theory, the rescue effect and the idea of propagule rain.

The Rescue Effect and Propagule Rain

An important assumption in the classical Levins model is that the extinction rate is linear with respect to the proportion of sites occupied (i.e., ep), which means that a constant proportion of occupied sites is expected to go extinct every time step. Hanski (1982) took exception to this assumption, reasoning that if p was high it would be likely that a subpopulation that was about to go extinct would be effectively "rescued" from its fate by migrations from nearby subpopulations. Thus, the extinction rate itself would be a function of proportion of sites occupied. Hanski proposed the model,

$$dp/dt = ip(1-p) - ep(1-p)$$

which can be rewritten,

$$dp/dt = (i-e)p(1-p)$$

which is the same as the simple logistic equation with the intrinsic rate of natural increase $i-e$ and the carrying capacity $K=1.0$. Obviously with an intrinsic rate of natural increase negative (which would be the case if extinction rate were greater than immigration rate) the overall population would tend to extinction (i.e., p would tend to zero), whereas if the immigration rate were greater than the extinction rate the overall population would tend to fully occupy all the sites (i.e., p would tend to 1.0). Thus, Hanski's model effectively predicts that there should be a bimodal distribution of populations when metapopulation dynamics are operative. Some populations should be near extinction while other populations should almost completely fill their potential habitats. This is referred to as the core-satellite hypothesis. The key here is that subpopulations shift between these two types, not that there are some species that are core and others satellite. Evidence for the hypothesis has been presented from a variety of sources (see figure 5.13).

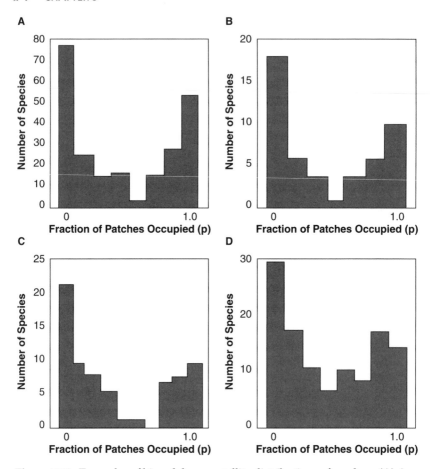

Figure 5.13. Examples of bimodal core-satellite distributions of p values. (A) Anthropochorous plants inhabiting small villages surrounded by forest (Linkola 1916; see also Hanski 1982). (B) British butterflies (data from Pollard et al. 1987; see analysis in Hanski et al. 1983). (C) Intestinal helminths in three species of grebe (pooled data for *Aechmophorus occidentalis, Podiceps griseigena,* and *P. nigricollis*; Stock and Holmes 1998). (D) Cynipid gall wasps on oaks (pooled data for *Quercus lobata, Q. chrysolepsis, Q. douglasii,* and *Q. egrifolia*; Cornell 1985).

Gotelli (1991) extended the model further by suggesting that the quadratic form of the immigration part of the original Levins model was too restrictive in that it presumed that propagules are more likely to arrive when more of the subpopulations are occupied. This scenario may be reasonable for some situations, but in others (e.g., plants with many and highly dispersable seeds) it does not seem to fit with reality.

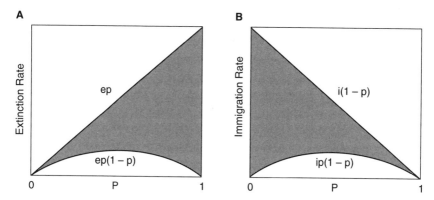

Figure 5.14. (A) Extinction rates in Levins's (ep) and Hanski's [ep $(1-p)$] models; p is the fraction of sites occupied. The difference between these two curves (dark region) constitutes the rescue effect. (B) Immigration rates in Levins's and Hanski's models [$ip(1-p)$] and Gotelli's model [$i(1-p)$]. The difference between these two curves (dark region) constitutes the propagule rain.

In many situations, there is likely to be a "rain" of propagules that is relatively independent of the number of habitats occupied; this characterizes a true island mainland situation, for example. Gotelli thus suggested modifying the immigration part of Levins's equation, obtaining the following model:

$$dp/dt = i(1-p) - ep(1-p)$$

where obviously the rescue effect has also been included. Here, the equilibrium value is

$$p^* = i/e$$

The rescue effect and the propagule rain are illustrated graphically in figure 5.14. The only other possibility left is for there to be no dependence on regional occurrence at all, namely,

$$dp/dt = i(1-p) - ep$$

which is Gotelli's immigration and Levins's extinction. All of these possibilities, along with their predictions of frequency of patch occupancy are presented in figure 5.15. Note that the only possibility that concurs with the data of figure 5.13 is when both immigration and extinction rates are dependent on regional occurence (i.e., the upper left in figure 5.15).

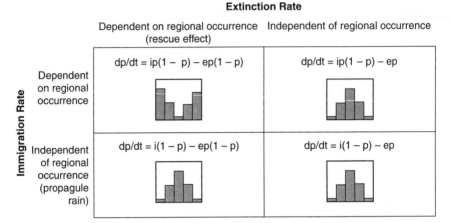

Figure 5.15. Summary of inclusion or exclusion of the rescue effect and propagule rain in the basic metapopulation model, including the expectation of distribution of patch occupancies for each model.

6
CHAPTER

Predator–Prey (Consumer–Resource) Interactions

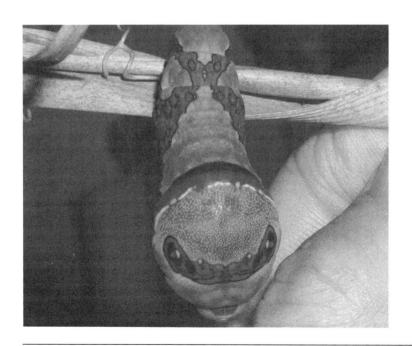

Chapters 1 through 5 dealt with the situation in which all interactions are among individuals within a single species. Individuals interact directly in order to reproduce, thus creating a birth rate, They interact indirectly when they use the same resources, thus creating the phenomenon of density dependence. They interact in complicated ways to create nonlinear effects, especially in structured populations. But all of that discussion involved individuals interacting in a single population or subpopulations interacting with one another in a larger population. Yet, in most real ecosystems critical interactions occur among species. A fir forest in northern Finland may be an excellent example of a population in which the dominant interactions are individual-to-individual within a single population of one species. But in a tropical rain forest in Borneo hundreds of species of trees interact with one another in a multispecies context. Here, it is population interacting with population in the context of competition among species of trees that is the critical force structuring the ecosystem. Even the simple density-dependent models we examined were implicitly cases of populations interacting with one another. When two individuals of the same species consume the same resource, if that resource is actually another organism, as it frequently is, the critical interaction is between the population of the consuming species and the population of the species that is consumed, a consumer–resource or predator–prey interaction.

Depending on one's intentions in developing a model it is frequently useful and certainly justifiable to take a phenomenological approach and simply model a single population as having a density-dependent

---

The evolution of elaborate snakelike patterns on this caterpillar suggests a defensive adaptation against predators, one of the most important forces of selection in nature.

effect without stipulating what is causing that effect. However, it is frequently the case that one needs to explicitly acknowledge the second species involved in the interaction and model the system accordingly. Thus, population ecology normally also includes the subject of populations of different species interacting with one another.

Although a classification of these interactions can get large and cumbersome (predator–prey, competition, mutualism, symbiosis, consumer–resource, herbivore–plant, facilitation, comensalism, and so on), from the point of view of developing basic theory there are really only three basic forms: First, when one species has a positive effect on a second species while the second species has a negative effect on the first (predator–prey, consumer–resource, herbivore–plant, parasite–host, facilitation–competition). Second, the two species have negative effects on one another (interspecific interference, interspecific competition). Third, the two species have positive effects on one another. We deal with each of these three forms, but not in a fairly distributed fashion. That is, as pointed out several times in the past (e.g., Risch and Boucher 1976), mutualism in its various forms is probably the dominant form of interaction in the world yet receives the least attention in ecology textbooks. The reason for such a bias eludes most ecologists. We follow the traditional orthodoxy of pretending that predation and competition are the dominant forms of interactions, or at least the ones most worthy of the development of theory. Consequently, we focus first on predator–prey interactions, separately analyzing relatively large things eating relatively small things (classical predator–prey theory) and very small things eating large things (classical epidemiology theory), and second on interspecific competition. This chapter and chapters 7 and 8 are organized along these lines. This chapter presents classical predator–prey theory; chapter 7 examines the basic ideas of epidemiology, and chapter 8 looks at classical ideas of interspecific competition and mutualism.

Predator–Prey Interactions: First Principles

One organism eating another organism, the elementary process of consumption, was the main subject of the theoretical studies of Lotka (1926) and Volterra (1926), the influence of which cannot be overestimated. In principle, these theoretical studies apply to a variety of situations, generally referred to as consumer–resource systems (including herbivores and plants, parasites and hosts, and so on). We begin with the simplest form of those theoretical studies.

Consider a prey, the population density of which is V (for victim), and a predator, the population density of which is P. Beginning with basic rules of exponential growth, we could write,

$$dV/dt = (b_1 - m_1)V \qquad (1a)$$

and

$$dP/dt = (b_2 - m_2)P \qquad (1b)$$

where b and m are birth rate and mortality rate, respectively (see chapter 1). However, by the nature of these two populations we can also say that the birth rate of the predator cannot be constant but must depend on the population density of the prey (we assume there are no alternative food sources for the predator, so if the prey population is zero, the predator's birth rate must also be zero). Furthermore, the mortality rate of the prey must depend on the population density of the predator (we assume there are no alternative sources of mortality other than being eaten by the predator). Assuming the simplest relations possible (as a first approximation), we can cast the birth rate of the predator as a positive linear function of the prey, or,

$$b_2 = r_1 V \qquad (2a)$$

The mortality rate of the prey can similarly be regarded as a simple positive linear function of the density of the predator, or,

$$m_1 = aP \qquad (2b)$$

Substituting equations 2a and 2b into 1a and 1b, we obtain,

$$dV/dt = b_1 V - aPV \qquad (3a)$$

and

$$dP/dt = r_1 VP - m_2 P \qquad (3b)$$

These are the classic predator–prey equations of Lotka and Volterra.

A rather different approach could be taken to the elementary theory of predator and prey interactions. Rather than thinking of equations, we begin the analysis by asking, What will be the general behavior of a predator and prey when they interact with one another? Let us presume that the predator will be able to increase its population only when there are a critical number of prey around. That is, if the prey population goes below some critical number, the predator population cannot maintain itself and begins declining. Similarly, we can postulate a critical number of predators, below which the prey population can "escape" predation to some extent and increase its population but above which the predator simply overwhelms it, causing it to decrease. These two critical points are illustrated in figure 6.1. Since these critical points separate the part of the space in which the population is increasing from that in which it is decreasing, the point itself must represent a population that is neither increasing nor decreasing. That point is referred to as an isocline. The isocline, formally, is the set of all points for which the derivative (with respect to time) is exactly zero, which is to say the set of points for which the tendency of the population to increase is exactly balanced by the tendency to decrease. We can put the

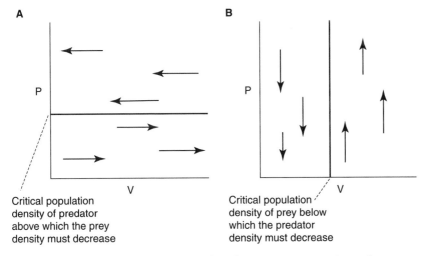

Critical population
density of predator
above which the prey
density must decrease

Critical population
density of prey below
which the predator
density must decrease

Figure 6.1. Qualitative construction of isoclines separating regions of space where (A) prey increases or decreases and (B) predator increases or decreases. P, predator; V, prey (victim).

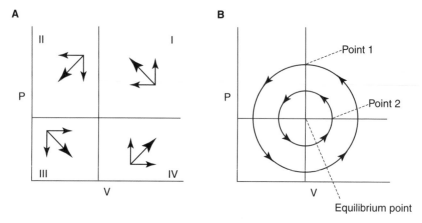

Figure 6.2. Construction of trajectories from isoclines. (A) The basic isocline structure leads to the vector field (large arrows). (B) Two illustrative trajectories.

two isoclines together and easily see the behavior of the two populations when they are interacting, as in figure 6.2.

In figure 6.2A the two isoclines separate four regions of the predator–prey space: one in which predator increases and prey decreases (quadrant I), one in which both predator and prey decrease (quadrant II), one in which predator decreases and prey increases (quadrant III), and one in which both predator and prey increase (quadrant IV). In figure 6.2B, two illustrative trajectories deduced from the simple vector field of figure 6.2a are presented. The most important conclusion to be drawn from this simple graphical analysis is that predator and prey will be oscillatory with respect to one another. This fact is intuitively obvious, at least with the help of the graph in figure 6.2A. This was the historically important fact about Lotka and Volterra's analysis.

Although the above derivation of the isoclines is intuitive, and certainly adequate, it is also possible to make the same derivation more formally. Returning to the formal model (equations 3a and 3b), we find the isoclines by setting the derivatives equal to zero, in which case we obtain,

$$P^* = b_1/a \tag{4a}$$

and

$$V^* = m_2/r_1 \qquad\qquad (4b)$$

which are the equations of the isoclines and correspond to the intuitive "critical population density of prey below which the predator must decrease" and "critical population density of predators above which the prey must decrease" (see figure 6.1).

The two-species equilibrium point is where the isoclines intercept: that is, where the two species are at equilibrium (see figure 6.2B). This equilibrium point is neither attractor nor repellor but is of neutral stability. Recalling the development in chapter 4, we can see that this situation is actually a bifurcation point and is structurally unstable, as will be apparent in a later section.

The most important prediction from the elementary theory is that predators and prey should oscillate with respect to one another, alternating between states of low predator–high prey numbers and low prey–high predator numbers. Does this prediction actually bear out in the natural world?

The most analyzed data set relating to this question is the lynx–hare data set accumulated over many years in Canada by the Hudson Bay company. On the basis of the number of pelts brought in each year and the assumption that the number of pelts hunters and trappers were able to obtain was correlated with the actual population density of the two species, it is possible to estimate numbers of individuals in the two populations over the general area traversed by hunters and trappers. These data have now been analyzed many times, frequently with great care and detail, revealing a variety of interesting details (for a recent analysis, see Blasius et al. 1999). But the most obvious result can be obtained from a quick glance at a graph of the data (see figure 6.3a). Predator and prey oscillate with respect to one another, just as the qualitative generalization of the theory shows.

Laboratory studies have been even more convincing. For example, Japanese workers counted numbers of Azuki bean weevils and their parasites over a period of several months (Utida 1957). The oscillations of predator and prey could not be clearer (figure 6.3B). In attempting to understand the dynamics of a parasite for the purposes of biological control, Huffaker set up a grid of oranges with complicated connections between them and followed the mite pest *Callosobruchus chinensis* and its mite predator *Heterospilus* over time (Huffaker 1958). His re-

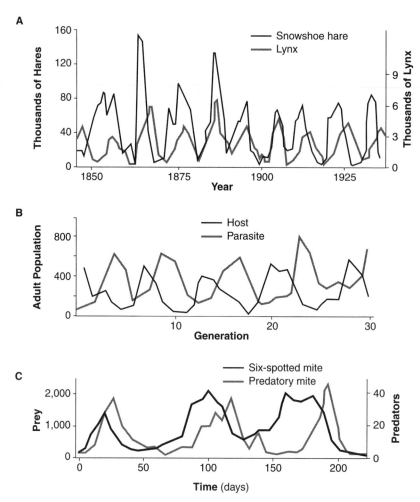

Figure 6.3. (A) Number of pelts of lynx and snowshoe hare brought into the Hudson Bay company by trappers. (B) Number of host Azuki bean weevils (*Callosobruchus chinensis*) and their parasite *Heterospilus* sp. (Utida 1957). (C) Number of six-spotted mites (*Eotetranychus sexmaculatus*) and their predator (*Typhlodromus occidentalis*) in the experiments of Huffaker (1958).

sults are displayed in figure 6.3C, again illustrating the qualitative conclusion that predator and prey tend to oscillate with respect to one another over time.

Density Dependence

The classic Lotka-Volterra predator–prey model leads to an important generalization, as discussed above: predator and prey ought to oscillate with respect to one another. However, it also results in a prediction that is universally agreed to be biologically absurd: whatever the mix of predator and prey, the system will always return to exactly that mix sometime in the future—the case of neutral stability. That states of neutral stability are useful points of reference in the global analysis of models was discussed in chapter 4, but here we seek to go beyond the simple assumptions of the Lotka-Volterra equations and add some measure of biological realism into the analysis of consumption, that is, predator–prey (consumer–resource) relations.

Recall the development of the logistic equation. We began with the process of exponential growth and proceeded by adding density dependence. The assumption that the per capita growth rate remained constant was replaced with the assumption that it varied negatively with the population density. It would seem obvious to do the same with the predator–prey situation. So, if we take the basic predator–prey equations,

$$dP/dt = rVP - mP \qquad (5a)$$

$$dVdt = bV - aVP \qquad (5b)$$

(the parameters have been changed and subscripts eliminated for clarity), using the same reasoning we used in deriving the logistic equation, it is not reasonable to assume that the per capita birth rate of the prey (b) is constant. This would be the same as presuming that the prey population could grow without limit if the predator was absent. This part of the equation can be made more biologically realistic by simply making the prey population obey the logistic equation, when the predator is absent. So equations 5a and 5b become

$$dP/dt = rVP - mP \tag{6a}$$

$$dVdt = bV[1 - (V/K)] - aVP \tag{6b}$$

where K is the carrying capacity of the prey population. The isocline for the prey equation (equation 6b) is

$$P = b/a - (b/aK)V$$

which is a linear equation in the P, V space. The qualitative dynamics that result from adding density dependence to the equation for V are deducible from an examination of the isoclines and how they change when the density dependence is added, as illustrated in figure 6.4. In the upper graph of figure 6.4 the original isocline is indicated by a dashed line, and the original vectors are also dashed.

The new isocline (the isocline that exists after density dependence is added to the prey) changes a piece of the region that used to be part of quadrant II or IV into part of what is effectively quadrant III or I (respectively). The change in vectors in this changed part of those quadrants is shown, whence it can be seen that a piece of the space now has vectors that point more toward the equilibrium point than they did with the original isocline. This difference causes the oscillations to move toward that equilibrium point. Thus, adding density dependence to the prey population causes the neutral oscillations to change into stable oscillations, which is to say the system is necessarily an oscillatory point attractor. This is not a trivial conclusion, for it restricts the behavior of predator–prey systems considerably, probably unrealistically so. It suggests that predator–prey systems are always stable! This is an unwarranted conclusion, as discussed in the following section.

Functional Response

The density-independent assumption for the prey species seems quite foolish, biologically. However, the parallel assumption for the predator is not all that unreasonable. It simply says that the predator will have a positive birth rate as long as there is food to eat, a statement that, under the restrictive set of assumptions used in developing these models, is sensible. However, there is another assumption about the predator

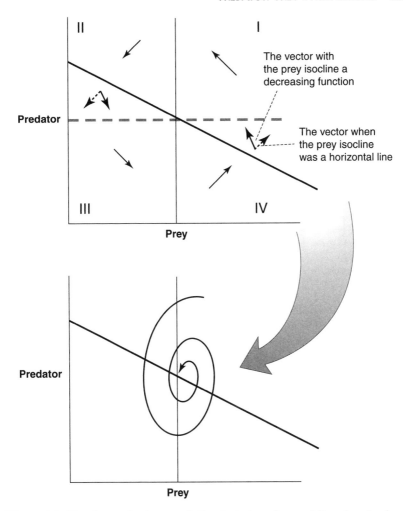

Figure 6.4. The change in the population trajectory from adding density dependence to *V*. Dashed lines are the original vectors and isocline.

behavior that is unreasonable. We have presumed that the ability of the predator to eat its prey is totally independent of the density of the prey, and similarly, the effect of the predator on the prey population is independent of the density of the prey population. These are the assumptions that we make in asserting that *r* and *a* are constants.

It is actually well documented (e.g., Hassell 1978) that prey are not eaten independently of the prey density. Indeed, for most predators, if

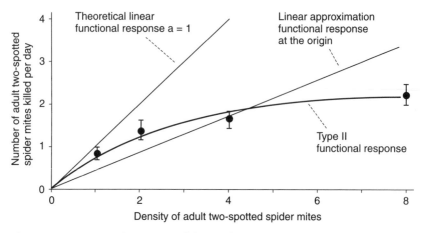

Figure 6.5. Functional response of the predatory mite *Neoseiulus barkeri* preying on the two-spotted spider mite (*Tetranychus utricae*) (from Fan and Petitt 1994).

you plot the rate of prey consumption against the population density of the prey, you do not get a straight line, as would be predicted by equations 6: that is, the predation rate is (aVP), which means that the predation rate is a linear function of the population density of the prey. For example, in figure 6.5 we see the results of an experiment with the predatory mite *Neoseiulus barkeri* preying on the herbivorous two-spotted spider mite *Tetranychus urticae* (Fan and Petitt 1994). On the other hand, such a functional response is not that mysterious. The original Lotka-Volterra equations presumed that predators will eat a certain fraction of prey, no matter how much is available. Thus, if a lion eats 1% of the zebra population each month, it will eat 1 of a population of 100, 10 of a population of 1,000, 10,000 of a population of 1,000,000. That is, it will never become satiated. This is obviously an unreasonable assumption for most predators.

Yet that is exactly the assumption made by the classic form of the Lotka-Volterra equations. In terms of actual data, the assumption is obviously ridiculous. For example, we have plotted the Lotka-Volterra assumption as a straight line with intercept at zero (that is, after all, what the assumption says) for the data in figure 6.5. In fact, what we actually see in almost all experiments is that the rate of predation is a curvilinear function of the population density, much as the function labeled Type II in figure 6.5, and a linear approximation is clearly unwarranted.

Because the predation rate is always a function of the population density of the prey, it is usually referred to as the "functional response." However, the exact form it takes can have a dramatic effect on the qualitative outcome of the predator–prey interaction. As we have already seen, if the functional response is linear, the system is either of neutral stability (if the prey are not density dependent) or approaches an oscillatory point attractor (if the prey are density dependent). What if the functional response is nonlinear?

This form of the function is most frequently modeled with the so-called Holling disk equation, after Holling (1959), which states,

$$\text{Predation rate} = \frac{cVP}{g + V}$$

an example of which is shown in figure 6.6A, labeled II. The meaning of the parameters in the Holling disk equation are illustrated in the figure. The value cP is the asymptotic value of the predation rate (assuming, for now, that P is unvarying), and g is the value that V assumes when the predation rate is at half its maximum value. Thus, c is proportional to the asymptotic value while g is a measure of the degree of curvature of the curve. Clearly, if $g = 0$, the functional response reverts

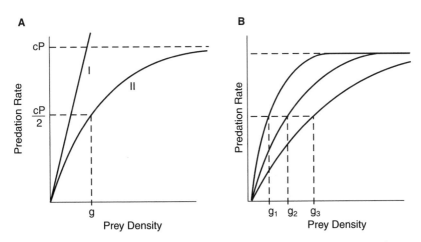

Figure 6.6. The functional response. (A) Linear form (I) and asymptotic form (II). (B) Three versions of a type II functional response with different values of the parameter g.

to its classical linear form, and as g increases, the curvature of the functional response slowly increases. Changes in the form of the functional response are illlustrated in figure 6.6B.

In addition to the strict mathematical meaning of the parameters c and g (which are made clear in figure 6.6), it is possible to ascribe more biological meaning to the parameters. Every predator needs to spend time finding prey. The time it takes to encounter an individual prey item will be an inverse function of the prey in the system (i.e., time to encounter a prey item $= a/V$), such that when there are large numbers of prey items, the time it takes to encounter one of them is very short, but when there are very few prey items, the time it takes to encounter one of them is long. But, after encountering a prey item, the predator needs to process that prey item (kill it, ingest it, digest it), commonly referred to as handling time. So the total time needed for a predatr to deal with a single prey item will be, $(a/V) + h$, where h is the handling time. The rate at which prey will be found and handled will then be the inverse of the time it takes to deal with an individual prey item, which is to say $1/[(a/V) + h]$, which we can rearrange to be, $(1/h)V/[(a/h) + V]$. Thus we see that the parameter c is the inverse of the handling time, and the parameter g is the ratio of the encounter time to the handling time.

The prey equation can be thought of as,

$dV/dt =$ rate of growth in absence of predator–predation rate

which is,

$$dV/dt = bV - aVP$$

when the predation rate is a linear function of the prey density. The rate of growth in the absence of the predator is bV, and the predation rate is $-aVP$. But this is assuming that the prey population is density independent (see the previous section) and that the predation rate is linear. If the predation rate is nonlinear, we can employ the disk equation of Holling to obtain,

$$\text{Predation rate} = \frac{cVP}{g + V}$$

which makes the prey equation,

$$\frac{dV}{dt} = bV - \frac{cVP}{g+V}$$

which has the isocline,

$$P = \frac{b}{c}(g+V)$$

Following the above argument, we must also modify the predator equation. That is, if the predation rate is an inverse function of the prey density, that fact must be represented in the predation rate of the predator equation also. That is, if the predator equation is

$$dP/dt = \text{birth rate} - \text{death rate}$$

and, as argued before, the birth rate is a function of the prey density, it is that birth rate that becomes satiated. That is, the birth rate cannot be a linear function of V, for the linear assumption implies that the predator can eat, effectively, an infinite amount of food. So the birth rate of the predator must be modified in the same way as the prey equation. Thus, if the original equation was,

$$\frac{dP}{dt} = rVP - mP$$

where rVP is the birth rate, we must modify that birth rate with the functional response. Thus the predator equation becomes,

$$\frac{dP}{dt} = \frac{rVP}{g+V} - mP$$

which has the isocline,

$$V = \frac{mg}{r-m}$$

which is, qualitatively, the same as without the functional response term: that is, a simple vertical line.

The two isoclines are plotted in figure 6.7, along with a qualitative interpretation of the vector field. Note that here we have a point repellor, of necessity. That is, according to the simple addition of the functional response to the classical Lotka-Volterra (LV) equations, we conclude that all predator–prey systems are unstable and thus cannot persist!

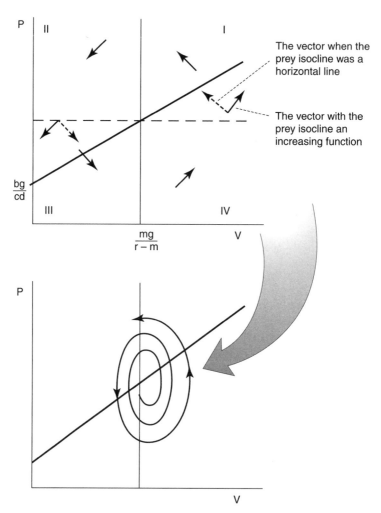

Figure 6.7. The change in the population trajectory from adding a type II functional response.

So, from two separate simple and biologically sensible modifications of the classical LV predator–prey equations we have: first, all predator–prey systems are stable (have oscillatory attractors); and second, all predator–prey systems are unstable (have oscillatory repellors).

Functional Response and Density Dependence Together

The truth of the matter is that the two nonlinear forces of functional response and density dependence seem to balance one another out. That is, the tendency of predatorprey systems to be unstable (resulting from functional response) is counterbalanced by their tendency to be stable (resulting from density dependence). This relation can be seen if we simultaneously add both nonlinear components to the original equations. Thus, for the prey equation we have,

$$\frac{dV}{dt} = bV\left(\frac{K-V}{K}\right) - \frac{cVP}{g+V}$$

The isocline for this equation is a bit more complicated than before, and is given as,

$$0 = bV\left(\frac{K-V}{K}\right) - \frac{cVP}{g+V}$$

which, after some algebraic manipulation, becomes,

$$P = \frac{b}{c}\left[g + \left(1 - \frac{g}{K}\right)V - \frac{1}{K}V^2\right] \qquad (7)$$

The important thing about equation 7 is that it is quadratic in the space of P,V, which means it is shaped like a parabola, as illustrated in figure 6.8a.

From a qualitative point of view, the key feature of this graph (figure 6.8) is that the prey isocline now may actually be ascending at the point where the predator isocline intersects. Looking at a close-up of

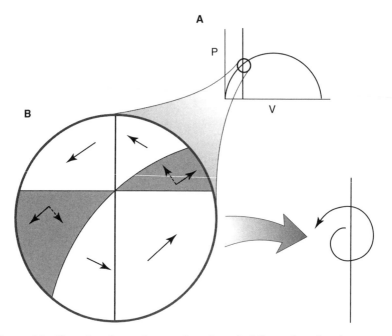

Figure 6.8. Changing dynamics as a function of adding a functional response to the Lotka-Volterra predator–prey equations.

that intersection (figure 6.8B) we see dynamic results that are similar to figure 6.7, except the isocline is not strictly linear. The change in the prey isocline from a horizontal line to an ascending function has caused the dynamics to destabilize, and the intersection point is actually a repellor. The predator and prey have expanding oscillations, coming ever closer to the origin, which is to say, ever closer to the extinction of the predator from the system. Because of the precise way in which the predator–prey equations are formulated it is actually not possible for the predator to become extinct, but it is possible for it to become "practically" extinct in that its numbers go so low that extinction through some random event is almost inevitable. As a rule of thumb, the closer the predator isocline is to the origin, the lower are the low points in the oscillations of the predator, as illustrated in figure 6.9.

Recalling the previous graphical analysis, one can easily see that if the predator isocline crosses the prey isocline to the right of the hump

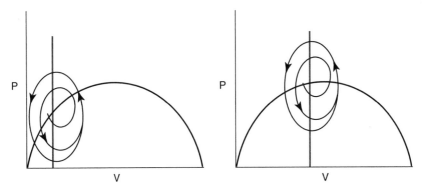

Figure 6.9. Trajectories of the predator–prey system with the predator isocline intersecting the prey isocline in its ascending limb.

of the prey isocline, the oscillations will dampen (i.e., the equilibrium point will be a focal point attractor). But also note that this powerful rule of thumb works only if the predator isocline is a simple straight vertical line.

Paradoxes in Applications of Predator–Prey Theory

Because of the inherent nonlinearities involved in predator–prey theory, there are sometimes unexpected consequences of particular actions that might be taken in management situations or that might result from evolutionary changes in parameters. Here we cite a couple of those so-called paradoxes.

First, consider the case of enrichment of the environment. This is something that is natural for a manager to try in a variety of situations. In a lake, for example, adding fertilizer would seem to be a good way to increase primary production; that increase would increase the carrying capacity of the zooplankton and would thus increase the catch of fish. Or so it would seem.

Recall the prey isocline,

$$P = \frac{b}{c}\left[g + \left(1 - \frac{g}{K}\right)V - \frac{1}{K}V^2 \right]$$

Calculate the peak of the hump of this function by first differentiating,

$$\frac{dP}{dV} = \frac{b}{c}\left[\left(1-\frac{g}{K}\right)-\frac{V}{K}\right]$$

and setting the derivative equal to zero and solving for V we have,

$$V^* = (K-g)$$

which we can substitute into the equation for the isocline to calculate the peak value of the isocline. Namely,

$$P = \frac{b}{c}\left[g+\left(1-\frac{g}{K}\right)(K-g)-\frac{1}{K}(K-g)^2\right]$$

which simplifies to

$$P = bg/c$$

Thus we can see these critical points on a state space graph, as in figure 6.10. Note that before enrichment, in this example, the system was a focal point attractor (i.e., a stable point); after enrichment proceeded to the isocline c (figure 6.10), the predator isocline intercepted the prey isocline to the left of its peak and thus the system was a focal point repellor (unstable). Thus, the paradox that enriching the environment has caused the system to shift from stability to instability (Rosenzweig 1971).

Concrete examples of the paradox of enrichment are not readily available in the literature, although there are some suggestive examples. For example, populations of the larch bud moth in Switzerland seem to exhibit cycles in the optimal part of its altitudinal range but are relatively constant in numbers in marginal (less "enriched") habitats (Baltensweiler 1971); this situation could be an example of the paradox (May 1981).

The paradox of biological control is another example. Biological control is effectively the application of predator–prey theory to the general problem of controlling pests in agriculture and forestry (Hawkins and Cornell 1999). Clearly, what is desired is the maintenance of the pest at

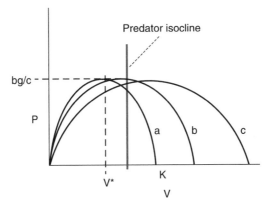

Figure 6.10. The paradox of enrichment. As the environment is enriched, the K of the prey species is increased, so that the prey isocline goes from a to b to c as the environment is enriched. However, the predator isocline remains constant.

as low a density as possible. Yet the elementary theory of predator–prey dynamics shows quite clearly that when the prey is at low density the system is very unstable (see figure 6.9), and it is likely that the biological control agent will be eliminated from the system (Arditi and Berryman 1991). So if the desire is to develop a system that will maintain itself indefinitely yet cause the prey to be at very low levels, we seem to have a paradox (see figure 6.11). Below, we discuss how this paradox was resolved in an actual case of biological control.

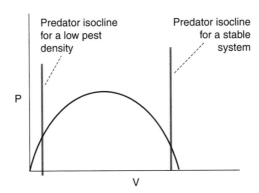

Figure 6.11. Isocline arrangements for the two goals of biological control, illustrating how the goals are clearly incompatible in theory.

Predator–Prey Dynamics: A Graphical Approach

The above development is based on an analysis of equations: the predator–prey equations with density dependence added to the prey and with the predator having a functional response of type II. It is useful to demonstrate how the same analysis can be heuristically presented in a simple graphical form, following the pioneering work of Rosenzweig and MacArthur (1963) and Noy-Meir (1975). Begin by plotting the rate of change of the prey population as a function of its own density. Note that this will *not* be an isocline but rather a plot of rate of change versus density (part of the whole idea is to derive the isocline from simpler first principles). At a very low density, the total population rate of change will be small, because there are so few individuals in the population. As the population density increases, so will the rate of increase, because there are more individuals reproducing. But we also expect that at high population densities the density-dependent effect will become important, and the rate of increase will begin to decline. Finally, at very high population densities, near the carrying capacity of the environment, the rate of population increase will approach zero. So, in summary, we can expect, qualitatively, (1) an increasing population growth rate as small values of population density increase, but (2) a decreasing population density as larger values of population density increase, with (3) zero population growth at zero population density and at the carrying capacity. The curve is indicated in figure 6.12, where it is labeled prey growth response. The curve in figure 6.12 is not an isocline. It is the result of plotting the rate of growth against the population density (not the population density of one species against that of another species).

We can also plot the functional response on the same graph, realizing that it is actually a negative rate of change. That is, the predator causes a decrease in the population of the prey. As we argued before, if there is a small number of prey items, even a single predator will eat all or most of them. If there are more prey items, the predator will eat more of them, but eventually the predator will become satiated. For example, let us suppose there is a fixed number of predators in the environment, say P^*. For that density of predators there will be a characteristic rate of depletion of prey, depending on the number of prey in the environment. If there is a small number of prey, we might expect all of them to be eaten. If the predators are not satiated, we could add yet

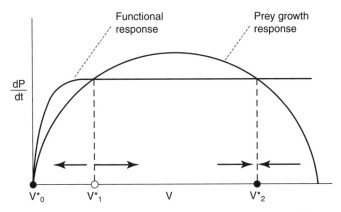

Figure 6.12. The two dynamic functions that make up the prey dynamic system.

more prey items to the environment and they would all, or almost all, be eaten. But ultimately we must be able to add so many prey items that all of the predators in the environment will be satiated. After that point, even if we supersaturate the environment with prey items, the rate of consumption of those prey items will remain constant, since the predators are satiated. Thus, we expect that a graph of predation rate versus V will be an ascending curve with diminishing returns, as illustrated in figure 6.12, where the curve is labeled functional response.

Examining figure 6.12, we can compare the prey growth rate with the functional response for particular values of V. If the growth rate is larger than the functional response, the overall population will increase (since the growth rate is the number of prey increasing per unit time and the functional response is the number of prey decreasing per unit time), and it will decrease whenever the functional response is greater than the growth rate. Thus, between the two points V^*_1 and V^*_2 in figure 6.12, the prey population will be increasing, as indicated by the arrows pointing to the right on the graph. However, to the right of V^*_2, the functional response is greater than the growth rate and we can expect the population to decline. Also, to the left of V^*_1 the prey growth rate is less than the functional response and we can expect the population to decline. The population declines are indicated by arrows pointing to the left in figure 6.12. A glance at this graph reveals the expected dynamics. V^*_2 is an attractor, as is V^*_0, while V^*_1 is a repellor. The point V^*_1 has also been referred to as a break point (May 1977) because it rep-

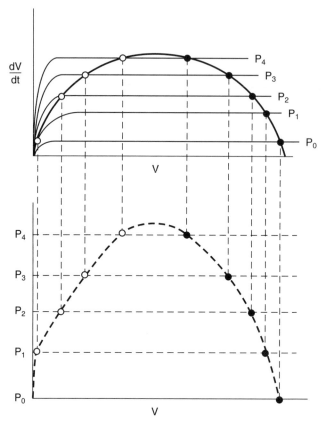

Figure 6.13. Constructing the prey isocline from the attractors and repellors of the rate versus density graph.

resents the value of the prey population density that can "break" a declining population and turn it into an increasing population.

This particular formulation is from Noy-Meir (1975) and was originally formulated in the context of grazing mammals, in which the predators are herbivores and the resources are plants. The formulation makes clear intuitively why the prey isocline has the shape it does. If we plot functional responses for different numbers of predators in the system, we expect the functional response to increase as the number of predators in the system increases, as illustrated in figure 6.13. This means that the exact location of the attractor and repellor (the intersection of the functional response curve with the prey growth rate curve) will change accordingly, as illustrated in figure 6.13. Plotting the attractors and repellors on a graph of V versus P, we obtain the points plotted on the lower graph in figure 6.13. It is clear that the prey isocline (where the prey is neither increasing nor decreasing, as defined before) appears with a "hump" on a graph of predator versus prey. Note that this curve is the same shape as the one derived analytically in equation 7 (the argument is actually quite the same, made more mathematically in leading up to equation 7 and more graphically here).

The basic shape of the prey isocline can be also deduced intuitively from reasoning about the isocline itself, as was done by Rosenzweig and MacArthur (1963). In the absence of a predator population, the prey population is expected to have some lower value below which it is unable to survive. For example, in the case of sexual animals, there is some low population density below which individuals of the opposite sex are unable to find one another very easily, and therefore the population will tend to decrease. At the other extreme, in a very dense population, the population will tend to decline, because it is above its carrying capacity. Thus we expect the sort of dynamic behavior illustrated in figure 6.14A. The middle point is a repellor, and both the upper point (the carrying capacity), and the lower point (extinction), are attractors. Now let us presume that a small number of predators are added to the system, but in such a way that they are nonreproductive and immortal; that is, we presume for temporary convenience that the predator population is unchanging. What will be the likely changes in the dynamics of the prey population? First, it is likely that the effective carrying capacity will be lowered, owing to the fact that the predators are eating a fixed number of prey animals each time period. Second, it is likely that the lower repellor will be higher; that is, the point at which the prey population is unable to sustain itself—to maintain its birth rate higher than its death rate—will be a larger value. It will require a larger inoculum of prey organisms to form a successful population. This situation is illustrated in figure 6.14B. If we now repeat this thought experiment of adding a fixed predator population to the system, we can trace out all the changes in both the key repellor and the carrying capacity, as is done in figure 6.14C. It is evident that a hump-shaped curve results, corresponding to the analysis of Noy-Meir as presented above.

The Rosenzweig-MacArthur approach can be utilized to model a variety of other qualitative situations, the most important of which is the

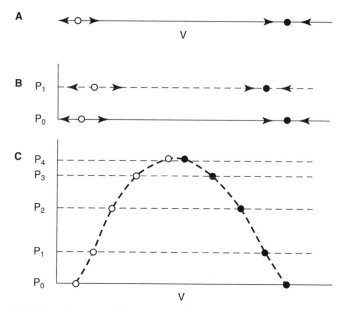

Figure 6.14. Construction of the prey isocline using the graphical arguments of Rosenzweig and MacArthur (1963).

addition of a refuge for the prey. If there is some part of the environment in which the prey population is invulnerable to the predator, there is a portion of the graph in which the prey is capable of increasing even though the predator population may be very large. Graphically, this situation is represented by a steep rise in the predator isocline at some low population density, specifically the population density that represents the size of the refuge (in terms of the number of prey items it will protect). The resulting qualitative appearance of the prey isocline is illustrated in figure 6.15, along with the expected dynamic behavior. Clearly, a refuge for the prey significantly "stabilizes" (in the sense that the limit cycle is constrained at a higher prey value than when the refuge was not there) the predator–prey oscillations.

A further complication is easily seen using the Noy-Meir approach. The type of functional response presented in figure 6.12 is probably the most common, but it is only one of three general types of response, as first noted by Holling (1959). The linear functional response simply proposes that the predator never becomes satiated. This may seem to be truly the case for some unwanted house guests, but, for the most part,

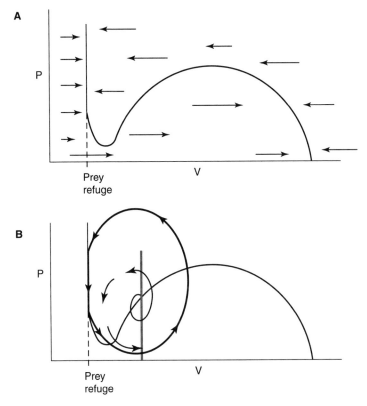

Figure 6.15. Prey isocline and its dynamics under conditions of a prey refuge. (A) The isocline. (B) The dynamic consequences.

consuming organisms get full, and they cannot ingest past some upper limit. Thus, most ecologists agree that the functional response must have some upper limit. But it is also sometimes the case that predators are able to fully exploit a prey item only after they have learned of its presence. And when a prey item is rare in the environment, there is less opportunity to learn of its presence as a potential food source than when it is common. Thus, the rate of predation increases very slowly at low prey densities but accelerates very rapidly after the predators have learned to eat the prey. We thus expect a differently shaped curve for a "learning" predator, one in which there is an accelerating rate at very low prey densities, followed by a decelerating rate, and ultimately, of course, by satiation. Thus the functional response curve should have a "sigmoid" shape, as indicated in figure 6.16 (III).

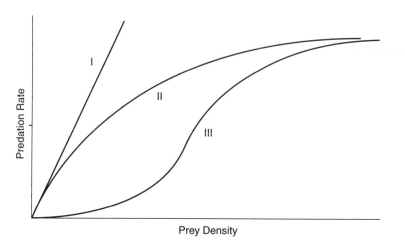

Figure 6.16. The three forms of the functional response.

Repeating the exercise that led to figure 6.13, the prey isocline is illustrated in figure 6.17 for the case of a type III functional response curve. Recalling the qualitative construction of the isocline with a prey refuge (figure 6.15), we see that a type III functional response has an effect that is similar to the addition of a prey refuge to the system.

The addition of a type III functional response, which in effect mimics the refuge addition in the graphical model, is a convenient way to theoretically resolve the paradox of biological control, as discussed above. If there is a refuge for the prey (or any of the other potential mechanisms that creates a prey isocline with a descending limb at low prey densities), the biological control paradox is resolved. The predator population can be retained in the system in a limit cycle with the prey item, because the latter is somehow protected from the predator when at low densities. For example, in Murdoch's classic studies of the California olive scale, careful estimation of parameters suggested that the system was unstable. Yet the parasitic wasp that was the biological control agent appeared to persist in the environment, maintaining the scales at a very low density. Careful observation revealed that scales accumulated on the trunks of the olive trees where the parasites could not find them, effectively a refuge in which a small population density

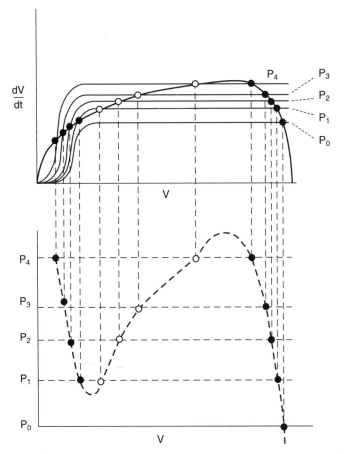

Figure 6.17. Constructing the prey isocline from the attractors and repellors of the rate versus density graph. The prey rate versus density graph has been distorted so as to illustrate the construction more clearly.

of scale insects could survive regardless the population density of the parasites (Murdoch et al. 1996).

Predator–Prey Interactions in Discrete Time

An alternative approach to predator–prey interactions is the well-known Nicholson-Baily model, which is effectively a discrete form of the Lotka-Volterra predator–prey equations. We here only briefly introduce the equations. Recalling the presentation of the exponential and logistic maps in chapter 5, we can begin the simple extension into the two-dimensional case with the simple exponential assumptions for each of the populations, namely,

$$V_{t+1} = bV_t \tag{8a}$$

$$P_{t+1} = cP_t \tag{8b}$$

But two independent exponential equations do not represent a good model of predator–prey interactions because they are not connected. The easiest conceptual way of linking these two equations is by defining a function, $f(V_t, P_t)$, that stipulates the probability that the prey will escape from predation during the time interval t to $t+1$ (Hassell 1978). Then equations 8 become,

$$V_{t+1} = bV_t f(V_t, P_t) \tag{9a}$$

$$P_{t+1} = cV_t[1 - f(V_t, P_t)] \tag{9b}$$

Thus equation 9a simply states that the number of prey individuals next time (V_{t+1}) is equal to the number now that escape predation $[V_t f(V_t, P_t)]$ times a reproductive factor (b), and equation 9b states that the number of predator individuals next time (P_{t+1}) is equal to the number of prey that succumb to predation now ($V_t[1 - f(V_t, P_t)]$) times the conversion of prey into predator individuals (c).

During a single time interval we expect that there will be a certain number of predatory events, N_e. That is, there will be N_e individual prey items consumed during the time interval. The crude attack rate (a) of predators then can be defined as the ratio of the number of prey attacked by an individual predator to the total number present, or,

$$a = N_e / V_t$$

The net attack rate is then the crude rate multiplied by the number of predators in the system. Thus we can write,

$$N_e = aV_t P_t \tag{10}$$

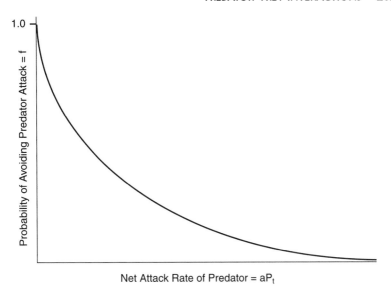

Figure 6.18. Probability of avoiding attack by the predator as a function of the net attack rate.

To avoid confusion we shall refer to the parameter a as the crude attack rate and aP_t as the net attack rate. What is likely to be the form of the function $f(N_t, P_t)$, the probability of avoiding predation? Certainly it must equal 1.0 when $P_t = 0$. Furthermore, as P_t approaches a very large number the likelihood of avoiding attack diminishes toward zero. Thus, we expect a relationship something like that presented in figure 6.18. Nicholson's original reasoning was simply to assume that attacks were random and thus the function f should be proportional to the zero term of the Poisson distribution (see chapter 5), namely,

$$f = e^{-(N_e/v_t)} \tag{11}$$

which is a convenient description of the general form shown in figure 6.18.

Equation 10 can be rearranged to give,

$$N_e/V_t = aP_t$$

and, substituting into equation 11, we have,

$$f = e^{-aP_t} \qquad (12)$$

Equation 12 can be substituted into equations 9 to obtain,

$$V_{t+1} = bV_t e^{-aP_t} \qquad (13a)$$

$$P_{t+1} = cV_t[1 - e^{-aP_t}] \qquad (13b)$$

which are the classical Nicholson-Bailey equations (Nicholson 1933, Nicholson and Bailey 1935).

7
C H A P T E R

Epidemiology

Chapter 6 treated the situation in which predator and prey are relatively equivalent in size and life history characteristics. The development is thought to apply to a wide variety of organisms, from parasitic hymenoptera attacking insect hosts to lions attacking zebras. But there is another kind of predation that is so different that it merits an entirely different mathematical approach. When the host is very large with relatively slow dynamics and the parasite is extremely small with very rapid dynamics the traditional predator–prey approach is not very useful. This is the case of infectious disease, in which the host is usually an animal or a plant and the predator is a microorganism such as a bacterium or a virus (referred to as the pathogen). In this situation, we usually assume that the host population is constant and ask questions about the spread of the parasite among host individuals. This is the subject matter of classical epidemiology, and we introduce it here in its simplest form. The interested reader is referred to one of the many excellent texts in epidemiology (e.g., Bailey 1975, Anderson and May 1991) for more sophisticated development.

Direct Disease Transmission

We begin by dividing the host population into individuals infected with the pathogen and those not infected (the susceptibles), in much the same way we divided the habitats into occupied and unoccupied

---

Every evening the egrets gather in this clump of trees, apparently to avoid the predators who are not enthusiastic to enter the swamp (which is filled with caimans). But such population congregations provide excellent opportunities for diseases to spread from bird to bird.

in the case of metapopulations (see chapter 5). Let I be the number of infected and S the number of susceptibles and suppose that $N=I+S$, where N remains constant as I and S vary through time. We presume the pathogen can exist only within a host and there are no intermediate hosts (which is the definition of a direct transmission system). The rate of increase of infecteds must be proportional to the product of infecteds and susceptibles (since the rate of transmission is proportional to the probability that susceptibles encounter infecteds). This is known as the mass action assumption. Thus we write,

$$\frac{dI}{dt} = aIS$$

Where the parameter a is the proportionality constant (usually referred to as the transmission coefficient) and essentially represents the probability that if an infected contacts a susceptible, the disease will in fact be transmitted. Since $N=I+S$, we know that $S=N-I$, so we can substitute for S in the above equation, to obtain,

$$\frac{dI}{dt} = aI(N-I)$$

And if we think in terms of proportions, let $N=1$ (so that I becomes the proportion of the total population infected), we have,

$$\frac{dI}{dt} = aI(1-I) \tag{1}$$

which is readily recognizable as the logistic equation, with the carrying capacity equal to unity.

Thus, in its simplest manifestation we expect the time course of a disease to have roughly a logistic pattern, which is what is frequently observed. For example, Stiven (1967) experimentally introduced a pathogenic microbe (*Hydramoeba hydroxena*) into a cniderian (*Chlorohydra viridissima*) and charted the time course of the infection. His results are illustrated in figure 7.1. Note the qualitative pattern, identical with the pattern predicted by the simple logistic equation (recall chapter 1).

As simple as this model is, and as nicely as some data fit the logistic

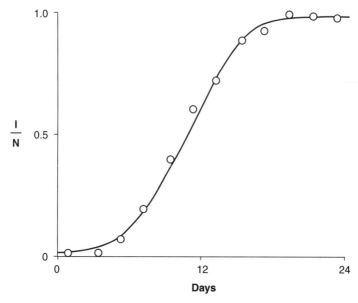

Figure 7.1. Time course of an infection. The proportion of a population of the cnidarian *Chlorohydra viridissima* infected with the microbe *Hydramoeba hydroxena* (from Stiven 1967 as reported in Anderson 1981).

pattern, most diseases are more complicated and include the concept of resistance or immunity. This is especially true when speaking of human diseases (or, for that matter, the diseases of most vertebrates, some invertebrates, and even some plants, depending on how one really defines immunity). Thus, rather than having a population structured with only susceptibles and infecteds, we really have three types: susceptibles, infecteds, and resistants (or recovered and immune). Those three types give rise to a basic model usually referred to as a SIR model.

Assume that the net rate of input of susceptibles exactly equals the rate of natural (non-disease-caused) mortality (bN) and disease-caused mortality (cI) (based on the assumption that the overall population remains constant). Thus,

$$\frac{dS}{dt} = bN + cI$$

represents the increase of susceptibles (where the parameter b is the rate of mortality for noninfected individuals and c is the rate of mortality for infected individuals). But if $bN + cI$ is the net rate of increase, the outflow from the population of S must include the rate of disease infection, aIS (see above) plus the rate of natural deaths of susceptibles, bS. Thus, the equation describing the overall dynamics of the susceptibles becomes,

$$\frac{dS}{dt} = bN + cI - aIS - bS \tag{2a}$$

The rate of change of infected individuals is the rate of increase (aIS) minus the rate of loss, which must be loss due to both natural (bI) and disease (cI) effects. In addition, some infected individuals will recover, so as to become resistant. Let the rate of recovery be v, so that the rate of loss of infecteds due to recovery is vI. The equation for infected individuals thus is,

$$\frac{dI}{dt} = aIS - (b + c + v)I \tag{2b}$$

The rate of change for recovered individuals is simply the rate of increase, which is the rate of recovery of infecteds (vI) minus the natural death rate, bR, which gives us,

$$\frac{dR}{dt} = vI - bR \tag{2c}$$

The equation system 2a, 2b, and 2c is the standard SIR model of direct disease transmission.

One of the most interesting conclusions to be drawn from this model comes from its equilibrium state. We are interested in the situation in which the equilibrium value of I is greater than zero, which is to say, the situation in which the disease will prosper. So, taking the equilibrium value of equation 2b, we obtain,

$$S^* = (b + c + v)/a$$

Note that $b + c + v$ is equal to the rate at which infected individuals are removed from the population. Since a is the transmission rate, we see that the number of susceptibles at equilibrium (S^*) is the ratio between the overall removal rate and the transmission rate. Now take the equilibrium value of equation 2a, obtaining,

$$I^* = \frac{bN - bS}{aS - c}$$

and, substituting the equilibrium value of S, and letting $b + c + v = \Psi$, we obtain,

$$I^* = \frac{bN - b\left(\dfrac{\Psi}{a}\right)}{b + v}$$

We are interested in the situation in which I^* is greater than zero, since this represents the situation in which the disease persists in the population. So setting $I^* > 0$ and solving, we obtain,

$$Na/\Psi > 1$$

The quantity Na/Ψ is referred to as the basic reproductive rate of the infection and is usually symbolized by R_0 (and pronounced R-zero). Thus we have,

$$R_0 = Na/\Psi$$

We now have a rather important generalization that is equivalent to ensuring that $I^* > 0$. If the basic reproductive rate is greater than unity, the disease will persist in the population, whereas if it is less than unity the disease will disappear. This quantity has been extremely important in understanding the qualitative nature of infections and, in some cases, public health planning.

For example, when planning a vaccination program, one presumes that the vaccination will change susceptibles into resistants. So both equations 2a and 2c must be modified to take into account the new rate. Letting the vaccination rate be i (immunization rate) from equation 2a we can write,

$$\frac{dS}{dt} = bN + cI - aIS - bS - iS \tag{3a}$$

$$\frac{dI}{dt} = aIS - (b + c + v)I \tag{3b}$$

and

$$\frac{dR}{dt} = vI + iS - bR \tag{3c}$$

where equation 3b is the same as equation 2b, but 3a and 3c have been modified to account for the vaccination rate i. As before, we calculate the equilibrium value of infecteds (set the derivative in equation 3a equal to zero and solve for I), whence we obtain,

$$I^* = \frac{bN - (b + i)S}{aS - c}$$

From this equation, it is clear that with any positive value of immunization rate, the equilibrium value of infections will decrease. Indeed, this is the experience in public health, not surprisingly. For example, the data for cases of individuals infected with measles in England and Wales for the years 1940 through 1984 are presented in figure 7.2. Note that the average number of cases, although highly variable, has a mean around 400 thousand reported cases per year from 1940 through 1966. Then, a vaccination program was introduced in 1966, and the reported cases dropped precipitously to about 100 thousand cases per year by the late 1970s and 1980s. This is precisely the pattern expected from the model presented in equations 3a, 3b, and 3c.

Note that the basic reproductive rate for this model is,

$$R_0 = abN/(b + i)\Psi$$

and we can ask the question, What value of i would be necessary to drive R_0 to unity (and thus extinguish the disease)? Setting $R_0 = 1.0$ and solving for i, we see that,

$$i = b(aN - \Psi)/\Psi$$

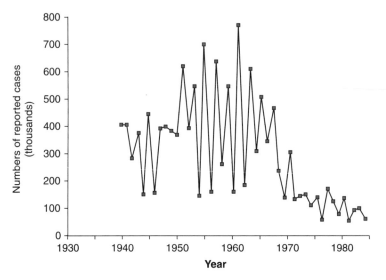

Figure 7.2. Number of reported cases of measles in England and Wales (data from Anderson and May 1991).

is the condition for which the disease will be extinguished, and we can see how such models are of help in designing immunization programs. If we know, or can guess, the parameters in this equation, we know what target vaccination rates ought to be.

Another important issue that emerges from this basic SIR model is the fact that the value of the basic reproductive rate is dependent on the total population size of the host (N). This means that R_0 is much more likely to be greater than unity in a large population than in a small population (all else being equal). It is also worth contemplating the meaning of the basic reproductive rate in a qualitative sense. It is the overall population density multiplied by the ratio of transmission rate and total rate at which infected individuals are removed from the population. Thus, if the transmission rate is very large, persistence of disease is ensured (not a surprising result). The basic reproductive rate is used as an indicator in planning vaccination campaigns. The point of the vaccination campaign is to reduce the value of R_0 below unity, which would cause the disease to disappear from the host population.

It is also notable that since the basic reproductive rate is dependent on N, the host population density, there will be a critical host population density below which the disease cannot exist. That critical density is,

$$N = \Psi R_0 / a$$

which says that the critical density for any disease is directly proportional to the basic reproductive rate. Qualitatively, we expect that in a growing host population, various diseases will become supportable at different times, depending on the basic reproductive rate. The higher the population density, the more diseases are expected to pass that threshold, a fact that has certainly been observed in the human population. It also places disease as a potential force in controlling populations. Indeed, one might even propose that it is possible to calculate the carrying capacity of a population on the basis of the basic reproductive rate of a controlling disease. This calculation would assume that the disease was the only population controlling factor and that it generates epidemics whenever it appears.

Indirect Disease Transmission

When another organism is involved in disease transmission, the basic dynamics of infection changes at a very fundamental level (Anderson and May 1979). Here we take the case of malaria as a model situation. Other vector-borne diseases are similar, but all have their own peculiarities and models must be constructed for each specific disease, although the underlying structure of the elementary malaria model has historically formed an important point of departure for the more complicated and realistic models. Furthermore, malaria affects more people worldwide than any other disease.

Ross (1911) was the first to treat this disease from a modeling perspective, and later Macdonald (1957) added considerable realism to the basic Ross model. All of this development is summarized by Anderson and May (1991).

We begin with some very simplifying assumptions. Suppose that there is no immune response in the host (now strongly suspected not to be true), that human victims of the disease do not suffer mortality but rather are eventually cured of the disease, and finally that the population density of both mosquitoes and humans is constant (i.e., the disease dynamics is very rapid compared with the population dynamics of either the host or the vector). The first of these assumptions allows us to formulate the problem in terms of two variables, the proportion of the humans that are infected (Y) and the proportion of the mosquitoes that are infected (X). The likelihood of a new infection is proportional to the probability that the mosquito is infected (X) times the probability that the human is not ($1 - Y$), and the rate of loss of old infections through recovery is r. Thus, the rate of change of the proportion of host individuals infected can be written,

$$\frac{dY}{dt} = aX(1-Y) - rY \tag{4}$$

where a is the proportionality constant. The parameter a is relatively complicated and depends on a variety of other parameters. Classically, it includes the biting rate of the mosquito (b), the proportion of bites that actually transmit the pathogen (c), the population density of the mosquito (N_y), and the population density of the humans (N_x). Note that the ratio N_y/N_x represents the number of mosquitoes per human host, which we can use because it is well known that an individual female mosquito has a constant biting rate—that is, a fixed number of blood meals per unit time—independent of the population density of the host. So we can postulate that the proportionality term, a, really should be equal to the biting rate times the proportion of bites that actually transmit the pathogen times the number of mosquitoes per human, which gives us,

$$\frac{dY}{dt} = (bcN_y / N_x)X(1-Y) - rY \tag{5a}$$

as the equation describing the dynamics of infection of hosts. Note that the term bcN_y/N_x is a single constant (originally a in the previous equation) written in terms of its biologically significant factors. The logic regarding the dynamics of the vector is similar in that the rate of acquisition of new infections should be proportional to the biting rate (b), the chance that a mosquito biting an infected host will acquire the disease (v) times the probability of encountering an infected host (Y) times the probability that the biting mosquito is not infected ($1 - X$). We must obviously subtract from this the death rate of the mosquitoes, uX. Thus we write, for the dynamics of the vector population,

$$\frac{dX}{dt} = bvY(1-X) - uX \qquad (5b)$$

Equations 5a and 5b are basically the same as the original model proposed by Ross (1911).

We analyze the model by looking at the isoclines, that is, where the derivatives are set equal to zero. The two zero isoclines are,

$$X = \frac{rY}{a(1-Y)}$$

for $dY/dt = 0$, and,

$$X = \frac{bvY}{u + bvY}$$

for $dX/dt = 0$. These two isoclines and their vector fields are pictured in figure 7.3. Putting these two isoclines together we can graphically analyze the system, as shown in figure 7.4.

There are two qualitatively distinct outcomes: one in which both populations form a stable equilibrium, the other where the disease disappears (number of infected humans and number of infected mosquitoes go to zero). It is of interest to note that when the disease persists (according to this model), the equilibrium state includes less than full

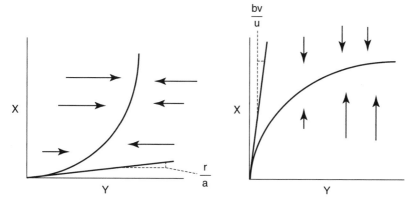

Figure 7.3. Isoclines and vector fields of system 5. The derivatives (slopes) of each of the isoclines evaluated at the zero point are also illustrated.

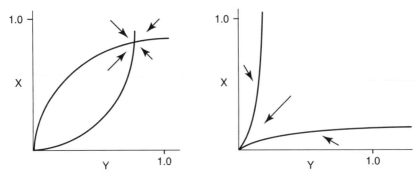

Figure 7.4. The two qualitatively distinct arrangements of the isoclines of system 5.

infection, which is to say, there will always be some humans and some mosquitoes that are not infected with the disease.

From an examination of the two graphs of figure 7.4, it is clear that the distinguishing feature of them is the way in which the slopes of the isoclines, evaluated at zero, are related to one another. By inspection of the graphs (and recalling the values of those slopes as shown in figure 7.4), we can say that the persistent case will exist when,

$$\frac{bv}{u} < \frac{r}{a}$$

If we let $a = bcN_y/N_x$, $m = N_y/N_x$, and rearrange a bit, we see that this condition becomes,

$$\frac{mb^2cv}{ru} < 1.0$$

In the classical malaria literature the left-hand side of this equation is usually symbolized as z_0 and is the basic reproductive rate of the disease. A bit of reflection will convince the reader that this measure is analogous to the basic reproductive value defined earlier for a directly transmitted disease. In the history of public health dealing with malaria, the basic reproductive rate has been an important measure. Strategies need to be devised to try and push z_0 as low as possible.

8
CHAPTER

Competition and a Little Bit of Mutualism

Competition: First Principles

In the nineteenth century, the notion of competition between different species of animals and plants was common. Indeed, competition was part of the intellectural context in which Darwin and Wallace formulated the theory of evolution through natural selection. But most frequently the idea of competition was thought of more or less as a sports metaphor. Two teams compete, one wins. The idea seems to have been too obvious to actually write about or think through too clearly. The insights of Lotka, Volterra, and Gause changed all of that.

At its most elementary level, interspecific competition involves two species utilizing a similar resource. It rapidly gets more complicated, but stripping the phenomenon of all its complications, this is the basic principle: two consumers consuming the same resource. If we suppose that the resource is a self-replicating dynamic one (we treat other cases later), we can treat it as a logistic population with two consumers utilizing it at two different rates. Thus, the resource equation is,

$$\frac{dR}{dt} = rR \left\{ \left(\frac{K-R}{K} \right) - a_1 N_1 - a_2 N_2 \right\} \tag{1}$$

where R is the density (or biomass) of the resource, r is its intrinisic rate of increase, K is the carrying capacity of the resource, a_i is the rate at which the resource is consumed by the ith consumer, and N_i is the population density of the ith consumer. Now we suppose that each of the

---

A butterfly picks up pollen from a tropical plant as it sucks up the nectar. The plant gives nectar to the butterfly, the butterfly helps the plant with sex, a case of mutualism, probably the most common interspecific interaction in nature.

consumer populations behaves according to the simple Lotka-Volterra predator–prey equations (see chapter 6), such that we have,

$$dN_1/dt = (b_1R - m_1)N_1 \qquad (2a)$$

$$dN_2/dt = (b_2R - m_2)N_2 \qquad (2b)$$

where the birth rate is proportional to the resource density such that b_i is the predation rate (consumption rate) of the ith consumer, and m_i is the death rate of the ith consumer. We now presume that the dynamics of the resource is very rapid compared with that of the consumer, such that we can set the derivative of equation 1 equal to zero and solve for the equilibrium value of R. After some algebraic manipulation we obtain,

$$R^* = K(1 - a_1 N_1 - a_2N_2)$$

Since we have assumed that the resource dynamics is very fast compared with that of the consumer dynamics, we can approximate the consumer dynamics at any particular value of N_1 and N_2, but simply substituting R^* in place of R in equations 2a and 2b. Thus, we have,

$$\frac{dN_1}{dt} = [b_1K(1 - a_1N_1 - a_2N_2) - m_1]N_1 \qquad (3a)$$

$$\frac{dN_2}{dt} = [b_2K(1 - a_1N_1 - a_2N_2) - m_2]N_1 \qquad (3b)$$

By rearranging, we can write,

$$\frac{dN_1}{dt} = (b_1K - m_1)\left(\frac{\dfrac{b_1k - m_1}{b_1a_1K} - N_1 - \dfrac{a_2}{a_1}N_2}{\dfrac{b_1K - m_1}{b_1a_1K}}\right)N_1$$

$$\frac{dN_2}{dt} = (b_2K - m_2)\left(\frac{\dfrac{b_2K - m_2}{b_2a_2K} - N_2 - \dfrac{a_1}{a_2}N_1}{\dfrac{b_2K - m_2}{b_2a_2K}}\right)N_2$$

Now make the following substitutions,

$$b_i K - m_i = r_i$$

$$\frac{b_i K - m_i}{b_i a_i K} = K_i$$

$$a_i / a_j = \alpha_{ij}$$

and we obtain,

$$\frac{dN_1}{dt} = r_1 N_1 \left[\frac{K_1 - N_1 - \alpha_{1,2} N_2}{K_1} \right] \tag{4a}$$

$$\frac{dN_2}{dt} = r_2 N_2 \left[\frac{K_2 - N_2 - \alpha_{2,1} N_1}{K_2} \right] \tag{4b}$$

which are the classic Lotka-Volterra competition equations. Note the term in brackets. It is very similar, indeed it is conceptually the same, as the density-dependent term in the logistic equation, except another species is involved. The alphas are referred to as the competition coefficients, and it can be seen here that they represent the utilization of the resource by one species compared with the utilization of that resource by the other species.

The equations above dealt with a single resource. Similar developments with two distinct resources become algebraically more cumbersome and are presented in a later section. These alternative developments are important in the sense that interspecific competition is almost never for a single resource (indeed, most ecologists think that two species competing for a single resource cannot persist together forever; that idea is described below). So the definition of the competition coefficient is not really as restrictive as it appears in the above development.

An alternative derivation of the classic equations follows along a more phenomenological path, based on the simple logistic equation. Namely, begin with the two equations

$$dN_1/dt = r_1 N_1 [(K_1 - N_1)/K_1] \tag{5a}$$

$$dN_2/dt = r_2 N_2 [(K_2 - N_2)/K_2] \tag{5b}$$

representing the logistic growth of two populations in isolation, where N_i is the population density of the ith species, r_i is the intrinsic rate of natural increase of the ith species, and K_i is the carrying capacity of the ith species. Recall the biological interpretation of the term $(K-N)/K$ as the proportion of environmental space (or critical resources, or places to hide). When another species is competing for that environmental space, the term must include that other species. Thus, rather than $(K-N)/K$, it is customary to write $(K_1-N_1-\alpha_{1,2}N_2)/K_1$, for species 1, and a parallel term for species 2. This term simply expands the general idea presented in the logistic equation, conceptualizing the second species as not quite equivalent to the first, which is what the alpha term, the competition coefficient, is all about. This coefficient effectively converts an individual of species 2 to an equivalent of an individual of species 1. For example, suppose the environment for species 1 includes 100 places to hide. Then $K_1 = 100$. Suppose that a single individual of species 2 needs one and one half of these places to hide. Then the value of $\alpha_{1,2}$ would be 1.5, indicating that two individuals of species 2 have the same effect on the environment as three individuals of species 1 (i.e., $3/2 = 1.5$). This stylized argument is presented pictorially in figure 8.1. Equations 5 are thus modified to become,

$$\frac{dN_1}{dt} = r_1 N_1 \left[\frac{K_1 - N_1 - \alpha_{1,2}N_2}{K_1} \right] \tag{6a}$$

$$\frac{dN_2}{dt} = r_2 N_2 \left[\frac{K_2 - N_2 - \alpha_{2,1}N_1}{K_2} \right] \tag{6b}$$

which are the classic Lotka-Volterra competition equations.

The traditional way of analyzing these equations is by examining the "isoclines" and deducing the qualitative behavior of the model (as we already have done with predator–prey models). The isocline is the set of all points for which the derivative is exactly zero, which is to say the set of points for which the tendency of the population to increase is exactly balanced by the tendency to decrease. Thus, setting the derivative of one of the equations equal to zero gives the equation for the isocline, which is to say, the equation for which the state variable in question does not change. So if we set the derivative of equation 6a equal to zero, we obtain,

$$0 = r_1 N_1 [(K_1 - N_1 - \alpha_{1,2}N_2)/K_1]$$

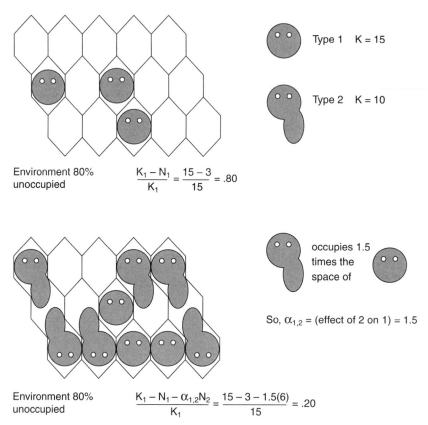

Figure 8.1. Diagrammatic representation of the meaning of carrying capacity and competition coefficient. The environment for organism type 1 contains 15 places. Organism type 2 occupies one and one half of the environment places that type 1 could occupy. So K for type 1 is 15, and K for type 2 is 10. Since type 2 occupies 1.5 of the environment space potentially occupied by type 1, the competition coefficient is 1.5.

Dividing both sides of the equation by $r_1 N_1 / K_1$, we obtain,

$$0 = K_1 - N_1 - \alpha_{1,2} N_2$$

which can be rearranged as,

$$N_1 = K_1 - \alpha_{1,2} N_2 \qquad (7)$$

which is a linear equation in the variables N_1 and N_2. If the above operations had been done with an inequality rather than an equality, we easily see that if,

$$dN_1/dt > 0$$

then

$$N_1 < K_1 - \alpha_{1,2}N_2 \tag{8}$$

Thus we have the important result that if relation 8 is true, the first population must be increasing. Using the same reasoning, we see that if,

$$dN_1/dt < 0$$

then

$$N_1 > K_1 - \alpha_{1,2}N_2 \tag{9}$$

If relation 9 is true, the first population must be decreasing. Examining relations 7, 8, and 9 on a graph (figure 8.2A) we immediately get an idea of the dynamic behavior (that is, changes through time) of the first population. Any population above the isocline will decline (relation 9), while any population below the isocline will increase (relation 8). Precisely the same analysis can be done for the second species, as presented in figure 8.2B.

The main interest in understanding competition is how the two populations interact when they are together. So it is necessary to put the two together by examining the dynamics represented in figure 8.2 on the same graph. In figure 8.3A the two graphs from figure 8.2 are su-

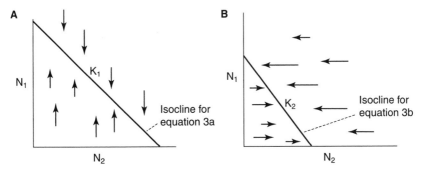

Figure 8.2. Isoclines of the competition equations.

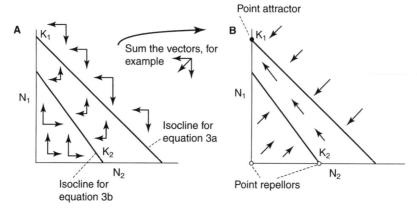

Figure 8.3. Putting both isoclines on the same graph and summing the vectors to get the vector field of the two-dimensional system.

perimposed on one another; the vectors (arrows) moving parallel with either the abscissa or the ordinate indicate the change of either one or the other species. But to examine the simultaneous changes in the two species, we must sum these two vectors: that is, determine the direction the two populations will go when simultaneously changing. This is done in figure 8.3B, with an illustration of how, qualitatively, the vectors are summed.

The diagram in figure 8.3B is a convenient tool for viewing the general dynamics of the system. It is relatively easy to imagine two populations beginning at an arbitrary point on the graph and being driven by the underlying biological rules in the direction indicated by the vectors. In this particular case, we see that the ultimate fate of competition is the extinction of species 2 and the total dominance of species 1. Analyzing the dynamics from a global perspective (that is, looking at the entire graph, not just the final equilibrium point), we can clearly see that the system has two repellors (one at K_2 and one at 0), and one attractor (at K_1).

In the classic analysis of competition, there are four general qualitative outcomes, based on the fact that you can place linear isoclines on this graph in four qualitatively distinct patterns, as shown in figure 8.4. Figure 8.4A is the same as figure 8.3B and has already been discussed. Figure 8.4B is the reverse situation of figure 8.4A, in which species 2 is the winner in competition. Thus figures 8.4A and 8.4B represent what

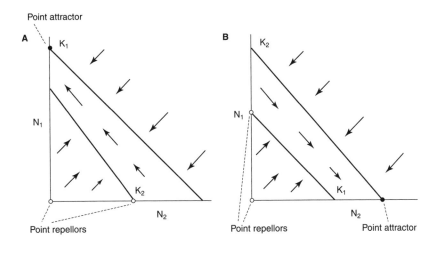

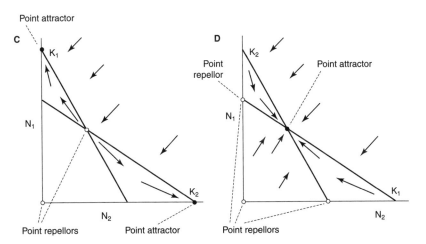

Figure 8.4. The four classic cases of interspecific competition.

some would intuitively think of as the competition process; either species 1 is the better competitor and wins or species 2 is the better competitor and wins. The sports metaphor applies well.

A slightly different situation is presented if the parameters are changed so that the two species are both strong competitors (the competition coefficients are large), as shown in figure 8.4C. Here, there will be a winner and a loser, but which species wins and which loses depends on where the ecosystem begins. Whichever species has the ad-

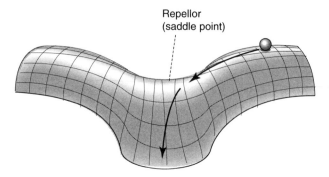

Figure 8.5. Physical model illustrating the behavior of a saddle point repellor.

vantage at the beginning will generally win in the end. Where the two isoclines cross is an equilibrium point, since, by definition, that is where the rate of change of the two species is exactly equal to zero. But an examination of the vector field (see figure 8.45C) shows that vectors move away from that point, meaning that it is a repellor. It is a slightly different form of repellor than we have seen before, since some vectors point toward it and other vectors point away from it. Rather than the physical model of a simple hill or valley (refer to the physical models in chapter 3), the appropriate physical model is a saddle, as we already discussed in chapter 4, and as illustrated here in figure 8.5. A marble balanced exactly in the middle of the saddle will stay there, unless perturbed, in which case it will fall off of the side of the saddle. It is also worth noting that this arrangement is highly sensitive to initial conditions (just as a chaotic system) in that the marble released from two apparently similar points may wind up at totally different attractors, which is to say very small changes in the point of initiation can have large effects in the future behavior of the trajectory.

One final situation, perhaps the most important one, is examined in the classic analysis of interspecific competition. This is the situation portrayed in figure 8.4D. An examination of the vector field shows that the equilibrium at which the two isoclines cross is an attractor. It is also the only case in which the attractor includes both species at positive densities, that is, the only attractor that does not involve the local extinction of one or the other species. This is the case of competitive coexistence.

An important distinction needs to be made between the graphs in figure 8.4A, B, and C (where one of the species is driven to local extinc-

tion), and the graph in figure 8.4D (where the two species exist together in perpetuity). In the first three cases, the outcome is similar to
that of the metaphorical sports match, where we expect a winner and a
loser. But the last case is effectively a tie. A better metaphor is a chess
match in which legitimate outcomes are either checkmate or stalemate,
the first when there is a clear winner, the last when the players have
played to the end and are tied. The difference between the two cases
forms the basis for one of the most important principles in ecology, the
competitive exclusion principle.

The competitive exclusion principle in its most elementary form has
quite a simple graphical interpretation. In figure 8.6 we illustrate the two
most important cases, the case of competitive exclusion in figure 8.6A
and of competitive coexistence in figure 8.6B. The equilibrium point occurs where the isoclines cross, as always. If a line is drawn between the
two carrying capacities (the dashed line between K_1 and K_2 in figure 8.6),
the position of the equilibrium point either above or below that line dif-
ferentiates between coexistence and exclusion. In figure 8.6A the point
falls below the line connecting the two carrying capacities, and exclusion
will take place. In figure 8.6B, the point falls above the line connecting
the two carrying capacities, and coexistence will take place.

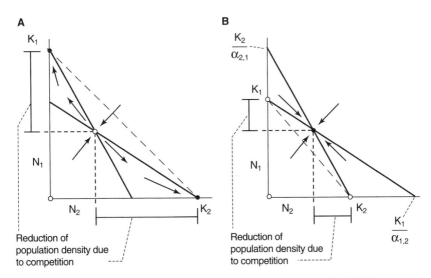

Figure 8.6. Explanation of the competitive exclusion principle. (A) Competitive exclusion. (B) Competitive coexistence.

The underlying reason for exclusion or coexistence can be readily seen on these graphs. The reduction from carrying capacity to the equilibrium point is illustrated on the two axes for both figure 8.6A and 8.6B. In the case of exclusion (figure 8.6A) the reduction has been relatively large, whereas in the case of coexistence (figure 8.6B), the reduction is relatively small. The principle, then, can be qualitatively stated that if competition is too large, exclusion will occur, and, most important, if competition is relatively weak, the two species can live together in perpetuity, despite the fact they are competing.

This idea is frequently related to the idea of the ecological niche, which itself has a long history in ecological thought. If two species have very similar niches (or very similar requirements for survival and reproduction), it is likely they will have large competition coefficients and extinction will therefore occur. If their niches are distinct, it is likely they will have small competition coefficients and coexistence will therefore occur. Fish do not compete with mice even though their ranges plotted on a map overlap. Their niches are obviously different. But fish may compete with tadpoles because both may eat the algae in the lake, although the tadpoles may eat smaller items than the fish and therefore their niches are not exactly alike. But if two mice eat exactly the same food, live in exactly the same habitat, have exactly the same nesting requirements, and so forth, they are likely to compete intensely, since their niches are so similar. One or the other species would likely be excluded from the region. Going from the fish–mouse competition (very low competition) to the fish–tadpole competition (intermediate competition) to the mouse–mouse competition, the competition coefficients go from zero to some high value, and the likelihood of competitive exclusion likewise goes from zero to some high value. To the extent the niches are different, the likelihood that exclusion will occur goes down, and to the extent the niches are the same, that likelihood goes up. Loosely speaking, this phenomenon is frequently summarized by noting that no two species can occupy the same niche (this point is treated more completely in a later section of this chapter).

More formally, note that the intercepts in figure 8.6 must be related to one another in a specific way to differentiate between exclusion (figure 8.6A) and coexistence (figure 8.6B). As indicated in figure 8.6B, the intercepts of the first isocline (the isocline for the first species) are K_1 (on the ordinate) and $K_1/\alpha_{1,2}$ on the abscissa. The intercepts of the second isocline are similar with appropriate adjustment in subscripts. The two graphs in figure 8.6 suggest that specific arrangement of the intercepts is required for coexistence (figure 8.6B). In particular, we require that the intercept of the first isocline on the abscissa be greater than the intercept of the second isocline on the abscissa, and the intercept of the second isocline on the ordinate be greater than the intercept of the first isocline on the ordinate. That is,

$$K_1/\alpha_{1,2} > K_2$$

and

$$K_2/\alpha_{2,1} > K_1$$

are the two conditions that must be satisfied if the two species are to coexist. With a small amount of algebraic manipulation we see that those conditions can be written,

$$\alpha_{1,2} < K_1/K_2 \tag{10a}$$

and

$$\alpha_{2,1} < K_2/K_1 \tag{10b}$$

A further bit of algebraic manipulation shows that a necessary (but not sufficient) condition for coexistence is,

$$1 - \alpha_{1,2}\alpha_{2,1} > 0 \tag{11}$$

Equations 10a and 10b are the formal conditions for competitive coexistence; equation 11 indicates the minimal condition on the competition coefficients themselves.

The competitive exclusion principle was an important point of departure for a great deal of ecological theory, especially in community ecology. To this day, in highly elaborated forms, the principle still forms the basis of much thinking about how ecological communities are structured. However, the specific form of the model (equations 6a and 6b) is too simple to reflect the complexities of the natural world. New models greatly complicate the entire field, and new ways of conceptualizing the competitive process have even challenged the underlying idea that no two species can occupy the same niche. Some of these newer approaches are discussed in a later section.

The Competitive Production Principle: Applications of Competition Theory to Agriculture

Intercropping or, more generally, multiple cropping (Vandermeer 1989), is frequently observed in traditional forms of agriculture, especially in the tropics. The basic idea is that more than one species of crop is grown in the same field, such that the two species are competing with one another, a clear case of interspecific competition.

Exactly what are the ecological benefits thought to accrue from the practice of multiple cropping? The hypothesized ecological benefits have been divided into two categories (Vandermeer 1989), reduced competition (or the competitive production principle) and facilitation. In the case of the competitive production principle, which we discuss here, it is thought that two species occupying the same space will utilize all the necessary resources more efficiently than a single species occupying that same space, thus corresponding to the coexistence criterion of the classic Lotka-Volterra competition equations, as described above.

The classical criterion for deciding whether an intercrop is better than its associated monocultures is the land equivalent ratio (LER) (which is equal to the relative yield total, RYT). If we have a total of 1 ha (say) and we wish to produce maize and beans, would it be better to produce them together as an intercrop or to divide the field into two parts and produce maize on one part and beans on the other? Presuming that the two crops compete with one another, we expect that the relative yield of each crop will be less than 1.0. That is, define the relative yield as P_i/M_i, where P_i is yield of crop i in the intercrop (P = polyculture) and M_i is its yield in monoculture; in both cases, the yield is expressed per unit area (kg/ha, or biomass/ha, or some other relevant measure). Then the relative yield total is simply the sum of the two relative yields,

$$RYT = LER = P_1/M_1 + P_2/M_2$$

The meaning of this measure is clear upon some reflection. If the polyculture produces some specific amount in a fixed area (say an area of 1 ha), how much area would be required to produce that same amount in two separate monocultures?

Thus, $LER > 1.0$ is usually taken as the criterion as to whether the intercrop will perform better than the separate monocultures. It is certainly not the only possible criterion (Vandermeer 1989) and sometimes it can be misleading, but it is normally taken as the first step in analyzing an intercrop. If LER is less than 1.0 the two monocultures would be better, but if LER is greater than 1.0 yield performance will be better in polyculture.

Since $LER = 1.0$ is the critical value deciding whether an intercrop will be better than separate monocultures, we can write,

$$1.0 = P_1/M_1 + P_2/M_2$$

which can be rearranged as,

$$P_1 = M_1 - [M_1/M_2]P_2 \tag{12}$$

If we plot equation 12 on a graph of P_1 versus P_2 we see that it is a line connecting the two monocultures (figure 8.7). Furthermore, we can easily see that $LER > 1.0$ represents the area above the line and $LER < 1.0$ represents the area below the line. Depending on the strength of competition (see explanation in figure 8.6), the system will be found either above or below the line connecting the two monocultures. The parallel between figures 8.6 and 8.7 is not accidental. The basic ecological process of competition is in operation in either case, and the question as to whether the intercrop will yield better than the two separate monocultures (figure 8.7) is parallel to the question as to whether the two species will coexist (figure 8.6). This parallelism between the classic ecological principle of competitive coexistence and polyculture advantage suggests that this form of polyculture advantage be termed competitive production (Vandermeer 1989).

Mutualism

A situation in which one population has a beneficial effect on a second population may be referred to as facilitation, a phenomenon that is becoming more frequently recognized in population biology. When two

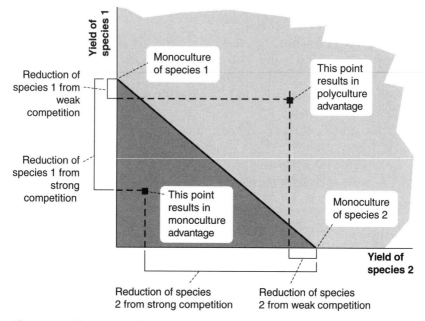

Figure 8.7. The graphical interpretation of the land equivalent ratio criterion (compare with figure 8.6). The darkly shaded area indicates monoculture advantage (below the critical line) and the lightly shaded area (above the critical line) indicates intercrop advantage.

populations simultaneously have beneficial effects on one another, the phenomenon is referred to as mutualism. Mutualisms are exceedingly important in the natural world.

Probably the simplest way of analyzing mutualistically interacting populations is with a simple modification of the Lotka-Volterra competition equations (Vandermeer and Boucher 1978). Modifying equations 6a and 6b, by making the competition coefficients mutualism coefficients, we can write,

$$\frac{dN_1}{dt} = r_1 N_1 \left[\frac{K_1 - N_1 + \alpha_{1,2} N_2}{K_1} \right] \tag{13a}$$

$$\frac{dN_2}{dt} = r_2 N_2 \left[\frac{K_2 - N_2 + \alpha_{2,1} N_1}{K_2} \right] \tag{13b}$$

which are identical to equations 6a and 6b except that the signs of the interaction coefficients are positive instead of negative. The isoclines are

$$N_1 = K_1 + \alpha_{1,2}N_2$$

and

$$N_2 = K_2 + \alpha_{2,1}N_1$$

graphs of which are shown in figure 8.8. In examining these graphs, we can see that several appear to make no topological sense (e.g., figure 8.8A) in that the vector field indicates a forever-increasing population. Clearly, this outcome is not possible. The purpose of this model is to establish the cases in which the mutualism will be maintained versus those in which it will not, and thus we are interested in general tendencies of the population behavior.

In cases in which the vector field indicates an ever-increasing population, obviously something else must happen, but that something else is beyond the analysis of the mutualistic system itself. So as not to leave this idea in the air, in all cases where the simple model indicates a forever-expanding population, isoclines have been extended with dashed curves, indicating the sort of qualitative changes that are likely to occur at higher densities.

There are eight cases. First we must distinguish between obligate mutualisms and facultative mutualisms. Facultative mutualisms exist when the two species are able to survive independently of each another but receive mutual facilitative effects from one another (figure 8.8A and 8.8B show the situation when both species are facultative). The famous mutualism between ants and the aphids they tend is such an example, in which both ants and aphids are capable of living independently, but both benefit from the presence of the other. Legume–*Rhizobium* symbioses are likewise usually cases of facultative mutualisms in that both legume and bacteria can live independently, but the benefit of living together is enormous.

Obligate mutualisms are indicated in this model with negative carrying capacities. The idea is that the very fact of obligateness indicates that the carrying capacity, the population density to which the popula-

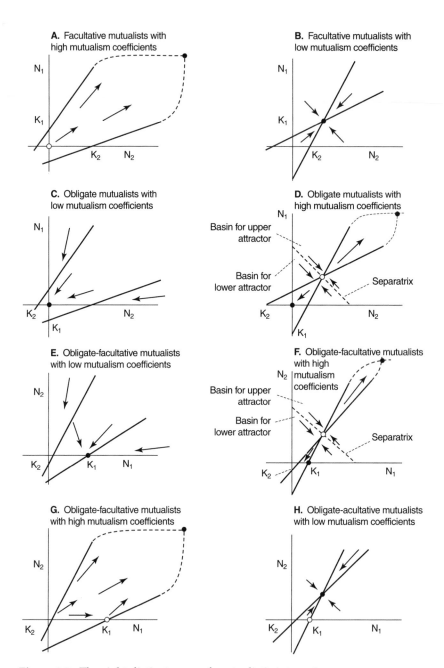

Figure 8.8. The eight distinct cases of mutualistic interactions.

tion would tend in the absence of interaction with the other species, cannot be positive. Formulating it as a negative value is obviously artificial but reflects well the idea of relative degree of obligateness: a large negative carrying capacity indicating a more obligate species than a small one. In figure 8.8C and 8.8D are the two cases in which both species are obligately mutualistic. If the mutualism coefficients are small (figure 8.8C) both species will go extinct, as would be intuitively expected (if both species are highly dependent on one another but provide very few benefits, those benefits are not enough to compensate for the extreme dependence). Obviously, one cannot come up with a natural example of such a state, but artificial ones are easy enough to manufacture. For example, if a VA mycorrhizal plant is planted and inoculated in a very acid soil with hardly any phosphorus, the plant may never be able to gain enough carbohydrate to transfer it to the fungus; if the plant can not enable the fungus to extract the phosphorus from the soil, both fungus and plant will die.

In figure 8.8D is the case of two obligate mutualists with high mutualism coefficients. Here, there are two basins: the upper is associated with the expanding population (expanding to some other point, not included in the model, as indicated by the dashed extensions of the isoclines), and the lower is associated with the extinction of both species. The separatrix (separation of the two basins of attraction) indicates critical values of the two species that must be attained so as to avoid extinction of both. The double obligate mutualisms that remain in nature tend to have very high mutualism coefficients (recall from figure 8.8C what happens if both are obligate with low coefficients). Perhaps the most spectacular examples are the mutualisms that are thought to represent the origin of many cell organelles (e.g., mitochondria and chloroplasts). Other examples abound: us and our bacterial gut flora, termites and their protozoan gut fauna, trees and mycorrhizal fungi, angiosperm pollination systems, and so forth.

Combining facultative with obligate mutualists generates parallel results, with slightly different specific outcomes, as shown in figure 8.8E,F,G,H. In all these cases, N_1 is taken as the facultative and N_2 as the obligate, and the results are parallel to the previous four cases if instead of the origin ($N_1 = 0$, $N_2 = 0$) being one of the equilibrium points, let K_1 be that point.

All of these cases and their qualitative outcomes, along with some examples of each, are summarized in table 8.1.

TABLE 8.1
Summary of Behavior of Mutualism, According to Equations 13a and 13b

First Species	Second Species	Level of Mutualism	K_1	System Behavior	Case in Fig. 8.8
Facultative	Facultative	High	>0	Expanding population	A
Facultative	Facultative	Low	>0	Attractor	B
Obligate	Obligate	Low	<0	Extinction of both species	C
Obligate	Obligate	High	<0	Alternative attractors (one at extinction of both populations)	D
Facultative	Obligate	Low	$>K_2/\alpha_{2,1}$	Extinction of obligate	E
Facultative	Obligate	High	$>K_2/\alpha_{2,1}$	Alternative attractors (one at extinction of obligate)	F
Facultative	Obligate	High	$<K_2/\alpha_{2,1}$	Expanding populations	G
Facultative	Obligate	Low	$<K_2/\alpha_{2,1}$	Attractor	H

Competition: The Details

The classic competition equations of Lotka and Volterrra may be classi-
fied as phenomenological in the sense that the process of competition
is the phenomenon of concern and it is modeled with a parameter that
represents that phenomenon directly (the competition coefficients). Although that is a useful first step in approaching the topic, the alternative mechanistic approach is the basis for most more-advanced treatments of the subject. With a mechanistic approach, ecologists focus on

how the consumption behavior of the competing organisms translates into the observed phenomenon of competition. Thus, rather than beginning with competition coefficients, we begin with the competing populations consuming resources (or otherwise modifying the environments of one another) and ask how the process of consumption translates into patterns of competition.

Mechanistic approaches to competition are most frequently based on Lotka-Volterra predator–prey equations, in one form or another. This basis is natural since we are concerned with how populations consume resources, which is structurally identical to how predators consume prey. The phenomena of birth and death and the fact that one population (the prey) promotes births in the other (the predator) while the other population (the predator) promotes deaths in the other (the prey) are usually regarded as primitive observables in this sort of mechanistic approach. A detailed philosophical analysis of the difference between phenomenological and mechanistic would undoubtedly reveal that subjective decisions on the part of ecologists have more to do with the distinction than any fundamental philosophical principles; one person's phenomenon is another person's mechanism. Yet, there is little question that the process of ingesting resources is a clear fact of nature, recognized as such by expert and nonexpert, while it is difficult to imagine picking up or directly seeing a competition coefficient. It is for this reason that the processes involved in the Lotka-Volterra predator–prey equations are usually regarded as basic, nonreducible features of the system, whereas competition coefficients are regarded as phenomenological representations. The coefficients of the predator–prey equations thus become the underlying mechanisms of the phenomenon of competition.

The first significant mechanistic approach to competition was that of Lotka, whose original derivation of the competition equations (equations 6) was actually based on the mechanistic ideas of two populations consuming resources. However, the most historically significant mechanistic approach was that of MacArthur, who had the insight to couple two predator–prey equations together to represent competition. We begin here with an even simpler approach and suppose that two consumers are utilizing a single resource. Using the basic form of the predator–prey equations and substituting R for N and C for P (emphasizing the term *resource* rather than simply population numbers and *consumer* rather than predator), the basic equation set is,

$$dR/dt = bR - aCR \tag{14a}$$

$$dC/dt = rRC - mC \tag{14b}$$

Following the previous analysis (equations 1–4), we allow the prey species to be density dependent, obtaining,

$$dC/dt = rRC - mC \tag{15a}$$

$$dR/dt = bR[1 - (R/K)] - aRC \tag{15b}$$

But here we are concerned with two species of predator (consumers), so we obtain,

$$dC_1/dt = b_1 RC_1 - m_1 C_1 \tag{16a}$$

$$dC_2/dt = b_2 RC_2 - m_2 C_2 \tag{16b}$$

$$dR/dt = b_3 R[1 - (R/K)] - a_1 RC_1 - a_2 RC_2 \tag{16c}$$

where the parameter r has been replaced with b_i, for notational convenience. Now we presume that the rate of change of the resource population is very large relative to that of the consumers so that we can allow the resource to reach its equilibrium. That is, the rates of change of the two consumer species are very rapid such that the resource population is constantly renewing itself and operating as if the two consumer populations were constant. In this case, we can set the resource at its equilibrium level and ask how the two consumer species will behave. Thus, we set equation 16c equal to zero, obtaining,

$$0 = b_3 R[1 - (R/K)] - a_1 RC_1 - a_2 RC_2$$

which we can rearrange to be,

$$K - a_1 KC_1 - a_2 KC_2$$

where R^* refers to the equilibrium value of R. Substituting the equilibrium value of the resource into the original consumer equations (equations 16a and 16b), we obtain,

$$\frac{dc_1}{dt} = b_1 C_1 (K - a_1 KC_1 - a_2 KC_2) - m_1 C_1$$

COMPETITION AND MUTUALISM   243

$$\frac{dc_2}{dt} = b_2 C_2 (K - a_1 K C_1 - a_2 K C_2) - m_2 C_2$$

with some algebraic manipulation we obtain,

$$\frac{dC_1}{dt} = (b_1 K - m_1) C_1 \left[\frac{\dfrac{(b_1 K - m_1)}{a_1 b_1 K} - C_1 - \dfrac{a_2}{a_1} C_2}{\dfrac{(b_1 K - m_1)}{a_1 b_1 K}} \right] \tag{17}$$

If we make the following substitutions,

$$r_1 = b_1 K - m_1 \tag{18}$$

$$K_1 = \frac{b_1 K - m_1}{a_1 b_1 K} \tag{19}$$

and

$$\alpha_{1,2} = a_2 / a_1 \tag{20}$$

equation 17 becomes,

$$dC_1/dt = r_1 C_1 (K_1 - C_1 - \alpha_{1,2} C_2)/K_1 \tag{21a}$$

and a similar analysis for the other consumer equation gives,

$$dC_2/dt = r_2 C_2 (K_2 - C_2 - \alpha_{2,1} C_1)/K_2 \tag{21b}$$

Equations 21a and 21b are identical to the Lotka-Volterra competition equations (equations 6a and 6b), whence we recall that the condition for coexistence of the competitors is,

$$K_1 / a_{1,2} > K_2$$

and

$$K_2 / a_{2,1} > K_1$$

We can substitute for the competition coefficients from equations 19 and 20, to obtain as the conditions for coexistence,

$$K_1 a_1 > K_2 a_2$$

and

$$K_2 a_2 > K_1 a_1$$

which obviously can never be true. This is the basis on which Lotka first articulated the competitive exclusion principle. Note that the underlying basis for this conclusion is that the consumer resource relationship is written in the form presented in equations 16, the most important assumption of which is a linear relationship between resource and consumer (i.e., the term $-a_1 RC_1 - a_2 RC_2$ in the resource equation and the terms $r_1 RC_1$ and $r_2 RC_2$ in the consumer equations); that is, the per capita rate of increase of either consumer or resource is related linearly to the other variable. Relaxing that assumption (Levins 1979, Armstrong and McGehee 1980) also eliminates the conclusion that two consumers cannot coexist on a single resource.

The important point about this formulation of competition is the meaning of the competition coefficients (and also the carrying capacity and intrinsic rate of natural increase for the consumers). The competition coefficient, α_{ij}, is clearly seen (equation 20) to be the ratio of the interspecific effect, a_1, to the intraspecific effect, a_2. This is the simplest mechanistic interpretation of the Lotka-Volterra competition coefficient.

MacArthur (1970) extended this approach to include two resources. Following the same development that led to equations 16, we can write,

$$dC_1/dt = b_1(R_1 + R_2)C_1 - m_1 C_1 \tag{22a}$$

$$dC_2/dt = b_2(R_1 + R_2)C_2 - m_2 C_2 \tag{22b}$$

$$dR_1/dt = b_3 R_1[1 - (R_1/K_1)] - a_{1,1}R_1 C_1 - h a_{1,2} R_1 C_2 \tag{22c}$$

$$dR_2/dt = b_4 R_2[1 - (R_2/K_2)] - h a_{2,1} R_2 C_2 - a_{2,2} R_2 C_2 \tag{22d}$$

Following the same procedure as the derivation of equations 21, we can convert equations 22 into the classic Lotka-Volterra form with the following substitutions,

$$K_1 = b_3 b_4 [b_1(K_1 + K_2) - m_1] / (b_4 K_1 a_{1,1} + h b_3 K_2 a_{2,1}) \qquad (23a)$$

$$K_2 = b_4 b_3 [b_2(K_2 + K_1) - m_2] / (b_3 K_2 a_{2,2} + h b_4 K_1 a_{1,2}) \qquad (23b)$$

$$r_1 = b_1(K_1 + hK_2) - m_1 \qquad (24a)$$

$$r_2 = b_2(K_2 + hK_1) - m_2 \qquad (24b)$$

$$\alpha_{1,2} = (b_4 K_1 a_{1,2} + h b_3 K_2 a_{2,2}) / (b_4 K_1 a_{1,1} + h b_3 K_2 a_{2,1}) \qquad (25a)$$

$$\alpha_{2,1} = (b_3 K_2 a_{2,1} + h b_4 K_1 a_{1,1}) / (b_3 K_2 a_{2,2} + h b_4 K_1 a_{1,2}) \qquad (25b)$$

and similar formulae for the second consumer species. Here, the mechanistic interpretation of the competition coefficients is somewhat more complicated but retains the essential symmetry of the ratio between the competitive effect of one species and that of the other species. For example, if the two resources are identical in their growth rates and carrying capacities (i.e., if $b_4 = b_3$ and $K_1 = K_2$), and if the consumers respond to the presence of each other identically (i.e., $h = 1.0$) , equation 25a becomes,

$$\alpha_{1,2} = (a_{1,2} + a_{2,2}) / (a_{1,1} + a_{2,1}) \qquad (26)$$

which is, of course, the competitive effect of species 2 divided by the competitive effect of species 1. Recall that a necessary condition for coexistence is,

$$\alpha_{1,2} \alpha_{2,1} < 1$$

However, if equations 25 are true and $h = 1.0$, then $\alpha_{1,2} \alpha_{2,1} = 1$, which means that coexistence is simply not possible with this system of equations unless the response of the competitors is distinct for the two resource species (i.e., $h \neq 1.0$).

We can also use a graphical approach to derive the conditions for coexistence of competitors using the same resources, as well as the traits that confer superior competitive ability when competitive exclusion occurs. The graphical approach was developed originally for a single resource and extended by Tilman (1982) to a second resource. This graphical approach also allows us to add a few more pieces of reality to the simple equations we already developed. The key point is to re-call that a competitive equilibrium in a mechanistic, or resource-based, model must have both the resources and the consumers being at equi-librium. As with the development of the equations in the previous sec-tion, we will start with how a single species uses a single resource, add a second species competing for the same resource, and then add a sec-ond resource.

For mathematical simplicity, the previous derivation assumed a lin-ear relationship between population growth rate and resources. How-ever, in reality, as a resource becomes more and more available, it is increasingly likely that different resources will become limiting. Therefore, growth as a function of only a single resource will eventu-ally asymptote (figure 8.9a) and the graphical model incorporates this shape, although this does not change any fundamental results. Note that the asymptote in figure 8.9A is for maximum per capita birth rate and not for population size (as in the typical graph of logistic growth; e.g., figure 1.9 or as in figure 8.9B). Population size is at an equilibrium when the birth rate is equal to the mortality. In figure 8.9A, mortality rate is shown as independent of resource availability, although this condition could be easily modified. The critical message of figure 8.9 is that when the consumer population is at equilibrium, the resource will also be at equilibrium; this amount is called R^* ("R star"). To see this, note on the graph that if R exceeds R^*, the birth rate (the solid line) ex-ceeds the death rate, so that population size increases. A larger popula-tion will then take up more resources, so fewer resources will be avail-able; that is, R declines back toward R^*. In contrast, if R is less than R^*, the death rate is greater than the birth rate. Therefore, the population will decline in size so that resource availability will increase back to-ward R^*.

Figure 8.9B shows the dynamics of the consumer–resource interac-tion. As C increases over time to an equilibrium (the asymptote in fig-ure 8.9B) R decreases from some initially high value (where no plants are taking it up) toward its equilibrium value at R^*. Where $C = C^*$, R is also constant at R^*. Note that R^* is a dynamic equilibrium just as C^* is. At C^*, individuals are still being born and dying, although the popula-tion size stays constant. Similarly, at R^*, R is still being supplied at

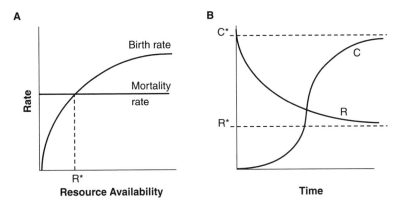

Figure 8.9. (A) Relationship between birth and mortality rates and resource availability, illustrating the meaning of R^*. (B) Growth patterns of the consumer (C) population and the resource (R) population over time.

some rate (called the supply rate), but that supply rate is equal to the consumption rate by the consumers.

Now add a second species competing for the same resource, for example, species B with a lower R^* and a lower maximum potential birth rate than species A (figure 8.10A). Is coexistence possible? Or, which species will be competitively superior and which will be excluded? We here present the graphical solution to these questions, but it is impor-

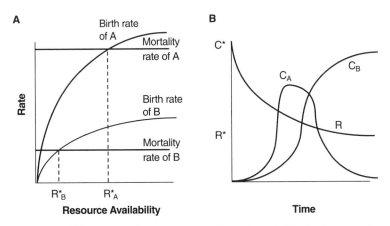

Figure 8.10. (A) Resource-dependent growth and mortality for two species using the same resource. (B) Time expected for both species and the resource.

tant to note that it has also been solved analytically by formulating equations for the birth and death functions in figure 8.10. Suppose initially R is quite high, above the R^* of both species A and B. In this case, both populations are growing (birth rate greater than death rate) and resource availability, R, will decline until it is below the R^* of species A. At this point, death rate exceeds birth rate for species A, and its population will decline. However, because species B is still growing, it will continue to deplete the resource even further, until R declines to R^* of species B. This analysis leads to the important and very general conclusion that under competition for a single limiting resource, the species with the lower R^* will always competitively exclude the species with the higher R^* (figure 8.10A). In other words, a species with a lower R^* can drive the resource to levels below that needed for inferior competitors to survive. As in the phenomenological Lotka-Volterra model or the simple resource competition model discussed above, two species cannot coexist on a single resource.

The dynamics of competition of two species for a single resource is shown in figure 8.10B. Note that the poorer competitor at equilibrium, species B with the higher R^*, initially increases faster because it has a higher maximum birth rate when resources are high (figure 8.10A) even though it is excluded at equilibrium when resources have become scarce. This result has important implications for successional trends; it also suggests caution in extrapolation of results from short-term competition experiments to the long-term outcome of competition.

What happens when a second resource is also potentially limiting? Is coexistence possible? Under what conditions? Tilman (1982) developed a graphical extension of the two species–one resource model to answer these questions. In figure 8.11 we plot the availability of resource 1 versus resource 2 and depict the combinations of the two resources that result in the consumer population's being at equilibrium, or the zero net growth isocline (ZNGI) for the consumer. If resources are substitutable, as are foods for most animals, greater availability of one resource lowers the requirement for another resource so that the ZNGI is a straight line with a negative slope (figure 8.11A). However, if resources are essential, increases in only one resource do not change the equilibrium requirement for a second resource. This requirement is the R^* for each resource so that the ZNGI for essential resources is right angled, with the elbow at the R^* for both resources (figure 8.11B). Tilman analyzed a number of different kinds of resources in his 1982

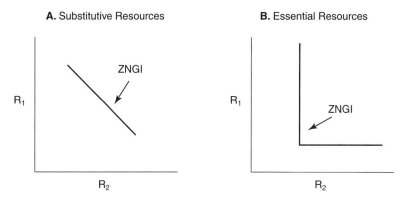

Figure 8.11. Zero net growth isoclines (ZNGI) for substitutive and essential resources.

monograph, but the model has been best developed for essential resources, and this is what is presented below. Essential resources are probably closest to reality for plants, which all require the same few resources (mineral nutrients such as N, P, K, and Ca and water and light; CO_2 and O_2 are also essential resources but are rarely limiting in terrestrial systems). Completely essential resources would mean that additional light, for example, could not compensate for less nitrogen. This is probably not strictly the case for plants; for example, higher light could enable the plant to assimilate more carbon and therefore to construct more roots to forage for nitrogen. This limited substitutability would result in a smoothing of the right angle of the ZNGI, but it would not change any of the conclusions developed below.

A given point on a ZNGI (all points at C^*) is also at an R^* if the consumption rate is equal to the supply rate (u). Although the real consumption rate of any plant is a complex phenomenon, it can be greatly simplified if we assume that the consumer takes up the two resources in exactly the ratio in which they are needed. This would be, in a sense, an optimal strategy for a plant because any additional amount of R_1, for example, could not be used to increase growth rate because there would be insufficient R_2 taken up. Therefore, the costs incurred in taking up the excess, or "luxury," amount of R_1 would be wasted. Because the ratio of the two resources required is the ratio of their R^*s, we can simply draw a vector from the elbow of the ZNGI (where $R_1 = R_1^*$ and $R_2 = R_2^*$) toward the origin to indicate consumption. The slope of this

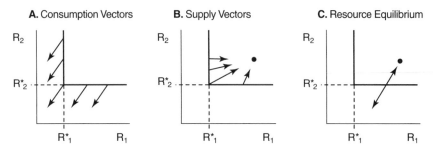

Figure 8.12. Consumption and supply vectors for the case of two essential resources.

vector indicates the ratio of the two R^*s consumed, and its length indicates the amount of each resource that is consumed. At the elbow, the ratio required is also the ratio available. However, if plants are truly optimal foragers, they should consume the resources in the ratio required regardless of availability. Therefore, the same consumption vector should apply anywhere on the ZNGI (or indeed on the entire phase plane in figure 8.12A).

The supply vector can also be constructed graphically given a simple assumption. The supply vector is the rate at which the resource is transferred from unavailable to available forms. For example, the nitrogen found in an ecosystem is unavailable when tied up in organic matter, whether in living plants and animals or in litter and soil particles. Available forms to plants are mineral nitrogen, either nitrate or ammonium. The transfer between living material and available forms is called mineralization and is facilitated by mircroorganisms. If we make the simple assumption that the supply rate is proportional to the proportion of the resource that is unavailable, we have the simple function: $u_j = a_j (S_j - R_j)$, where S_j and R_j are the total (supply point) and available amounts of resource j, a_j is a rate constant, and u_j is the supply rate. This assumption allows a simple graphical trick: starting from any combination of R_1 and R_2 on the phase plane, the supply vector will always point to the supply point (figure 8.12B). If, for example, most of resource 1 is already in available form (high R relative to S), little new resource can be supplied and much more of resource 2 will be supplied than resource 1. The length of the supply vector is also determined by the supply point. When the total supply of a resource is much higher than the amount available, supply rate will be higher (the vector longer) than when most of the resource is already in available form.

Combining the consumption and supply vectors, you can see that whether a given point on a ZNGI is both a consumer and a resource equilibrium depends on the supply point (figure 8.12C). For the resource to be at equilibrium, the slope and length of the supply and consumption vectors need to match exactly. Thus, for any given supply point, and assuming a constant and optimal consumption vector, there is only a single combination of resource availabilities on the ZNGI that is an equilibrium for both consumers and resources.

The final step in addressing questions about the outcome of competition for two species competing for two resources is to add a second species by adding a second ZNGI. Four possible ways of combining two ZNGIs exist (figure 8.13). In cases 1 and 2, the two ZNGIs do not cross, and therefore there is no two-species equilibrium point and no

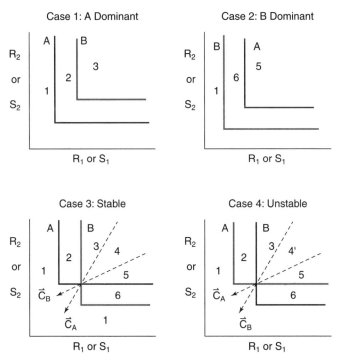

Figure 8.13. The four distinct cases of resource competition. Numbers 1–6 designate habitats.

coexistence. In case 1, the ZNGI for species B lies inside that of species A; that is, species A has a lower R^* for both resource 1 and resource 2. If the supply point is below both ZNGIs, neither species can survive regardless of competition. If the supply point is between the ZNGIs, resources are too low for species B to survive regardless of competition but sufficient for species A. If the supply point is above both ZNGIs, species A will eventually deplete both resources to levels lower than the minimum required for species B to balance its mortality rate by its birth rate. Thus, regardless of supply point, species A will always competitively exclude species B. Case 2 is identical except that the two species are reversed.

In contrast, cases 3 and 4 both have the ZNGIs crossing and therefore a combination of availability of resource 1 and 2 for which both consumers have stable population size. Therefore, coexistence is possible, although it turns out that the equilibrium is stable in case 3 but not in case 4. For the ZNGIs to cross, it is apparent that a tradeoff in competitive ability for the two resources is necessary: the better competitor for resource 1 (lower R_1^*) must be an inferior competitor for resource 2 (higher R_2^*) and vice versa.

Because the outcome of competition depends on the supply point, we will ask for which range of supply points the two-species equilibrium point is stable. To answer this question, we need to construct the joint consumption vector and then find the supply vector that exactly balances this consumption. The exact consumption vector depends on the relative abundance of the two species, but we know it must lie between the consumption vector of a monoculture of species A and that of a monoculture of species B. In case 3, if we extend these backward to find supply vectors that balance these consumption vectors, we enclose a triangular area of possible supply points for which coexistence is possible (habitat 4 in case 3 of figure 8.13). What if the supply point is not in this "coexistence region"? In habitat 1, resources are insufficient for both species regardless of competition, while resource 1 is insufficient for species B in habitat 2 and resource 2 is insufficient for species A in habitat 6. Habitats 3 and 5 are the most interesting because, in the absence of competition, both species could persist if supply points were in these regions. By examining what happens in these regions, we gain some clues about exactly why coexistence is possible in habitat 4 (figure 8.14). Suppose we have a supply point at X in habitat 3. If only species A or only species B were present, the equilibrium

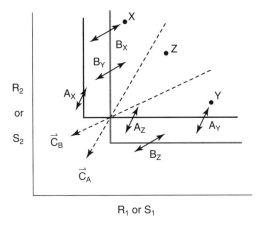

Figure 8.14. All possible equilibrium configurations in competition for essential resources.

would be at point A_X or B_X, respectively. Therefore, both species are limited by resource 1 (on the vertical part of their ZNGIs), and, because species A is a better competitor for resource 1, it will drive resource 1 to levels below the minimum required for species B. The opposite happens if the supply point is in habitat 5, for example, at point Y. In this case, both species are limited by resource 2, and species B, the better competitor for resource 2, competitively excludes species A. Now try the same exercise with the supply point Z in habitat 4, where the two species do coexist. In this case, species B is limited by resource 1, but species A is limited by resource 2.

This result gives us two important criteria for stable coexistence. First, the two species must be limited by different resources. Second, each species must be limited by the resource for which it is a poorer competitor. Species A is a better competitor for resource 1 but is limited by resource 2 within the habitats in which coexistence is possible. Species B is a better competitor for resource 2 but is limited by resource 1 within the habitats of coexistence. An intuitive explanation for this second criterion is that if a species were limited by the resource for which it is a better competitor, it would reduce the resource to its R^*, which, by definition, would be below that of the poorer competitor and therefore no coexistence would be possible. Because optimal foragers always consume relatively more of the resource for which they are a poorer competitor (i.e., for which they have a higher requirement), it turns out that as long as both species are optimal foragers, these two criteria will always hold. Therefore, as long as there is a tradeoff in competitive ability among species for different resources, the two additional criteria for coexistence developed here should always hold for some possible supply points. Case 4 in figure 8.13 shows an example identical to that of case 3 except that neither species is an optimal forager and each species consumes relatively more of the resource for which it is a superior competitor. In this case, despite the existence of a two-species equilibrium point, the equilibrium is not stable.

To summarize, this graphical model of resource competition comes to essentially the same conclusions as the simpler, phenomenological Lotka-Volterra competition models—species must differ in resource use to coexist—but it provides a much more detailed understanding of the mechanisms of interaction and coexistence and how the interaction depends on environmental conditions. For example, one simple insight from this model provides a potent critique of a common interpretation of field data on distribution patterns. Suppose you saw that abundance of a particular plant species increased with the availability of nitrogen in the soil. The intuitive interpretation and hypothesis based on this correlation would be that nitrogen must be a limiting resource for that species and that the higher nitrogen is responsible for the higher plant abundance. However, the resource competition model suggests exactly the opposite interpretation! If a resource is limiting (and a population is at equilibrium—an important caveat for this alternative interpretation), then the population should drive the availability of that resource down to the same R^* regardless of the total supply of that resource in the environment (i.e., the supply point). Therefore, a positive correlation between population size and any resource should actually be regarded as evidence that that resource is *not* limiting because it is not being consumed down to its R^*, and the positive correlation must be to some other factor (e.g., supply rates of different resources—a nonlimiting and a limiting one—may be correlated along a gradient because changes in moisture and temperature allow microbes to process both resources faster).

9
C H A P T E R

What This Book Was About

The subject matter presented in this text is logically organized, at least in the minds of its authors. In figure 9.1 we illustrate our vision of how all of the chapters are related to one another.

At the base is the subject matter of chapter 1, the central ideas of exponential growth and density dependence (one example of which is the logistic equation). From that base, we branched out into a specific application, the understanding of life history phenomena, as an example of the utility of the models presented in chapter 1. Then we moved into two general complicating issues, adding structure to the basic models (chapter 3) and the consequences of nonlinearities in general (chapter 4). We then looked to extensions of the basic models, in two distinct directions. First, we gave an extremely elementary introduction to the ideas associated with space, in chapter 5, including questions of both pattern analysis and the related issue of metapopulations. Finally, we presented three issues concerned with species interactions: predator–prey interactions (chapter 6), microparasites (epidemiology; chapter 7) and competition (chapter 8). Each of the tips of the branches on this tree (figure 9.1) could be extended into more complicated analysis. This text is intended as a springboard for all of those more sophisticated analyses.

It is also useful to say something about what is not in this text. Part of the science of ecology stems directly from the material presented here. The basic question that drives much of ecological thought is the question of species diversity; for example, Why are there more than 200 species of trees in a hectare of tropical rain forest but only three in a

255

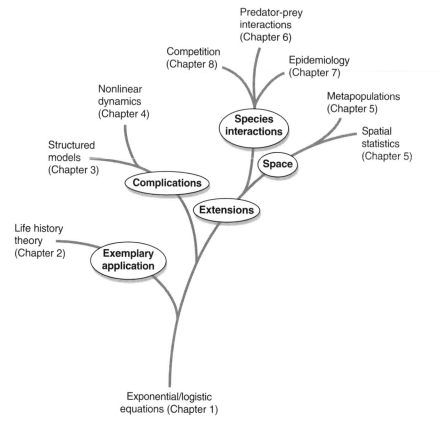

Figure 9.1. Diagrammatic representation of the subject matter of the text.

hectare of boreal forest? This is, in its broadest outlines, the subject of community ecology. It does not take much thought to extrapolate from the two species in competition of chapter 8 to the multiple species in competition in a community of grass species in a savanna. Indeed, an analysis of the community matrix, introduced for a two-species system in chapter 8 but analyzed in the literature for a multiple-species system, forms the basis of much classical community analysis (e.g., Levins 1968, May 1973). An alternative, but related, approach begins with the predation equations of chapter 6 and constructs complicated food webs on the basis of their structure (e.g., Hastings and Powell 1991).

But community ecology is only the most obvious extension of the material in this book. Chapter 7 introduces the analysis of microparasites, which requires a distinct break with the population dynamics of exponential and logistic equations and uses the metapopulation (with individual animals or plants as the habitat patches and the disease organism as the occupant of those patches) as the underlying theoretical framework. This is the subject of epidemiology, which has its own literature, to say nothing of its own departments in most major universities. And here the literature is truly staggering, with a history predating most of the rest of population ecology (indeed epidemiological studies of schistosomiasis alone far outnumber and predate most other approaches in population ecology, with one reference [Warren and Newill 1967] listing over 10,000 papers on the subject). Yet, we think it not excessive hubris to claim population ecology as the fundamental organizing principle of epidemiology. Indeed, one of the most highly respected treatises on human disease was written by two ecologists (Anderson and May 1991). Furthermore, a plausible argument can be made that an ecological approach to human disease is the wave of the future (or it should be).

A third field that emerges from population ecology is the study of spatial patterns in biological populations and communities. This field, which has burgeoned since the popularization of GIS (geographic information systems) technology, is based on the elementary introductory material supplied in chapter 5. That chapter introduces not only the idea of clumping in space but also the idea of spatial scale, which has become exceedingly important in recent work (e.g., Pascual and Levin 1999, Pascual et al. 2001). Moreover, we used the fact of spatial clumping to introduce the idea of metapopulations, a subject that has seen an enormous literature in recent years (summarized in Hanski, 1999).

Mastery of the material in this book will enable the student to delve into most of the literature in these more advanced topics with a sound analytical base. However, there is a perhaps more important use for the material presented in this text. Many of the world's most pressing problems will be solved only through the application of this material. In much the same way that physics and chemistry supplied the underlying scientific basis for engineering during the late phases of the Industrial Revolution, and similar to the way in which molecular biology is emerging as the underlying scientific basis for the development of new pharmaceuticals, population ecology is the scientific base upon which solutions to many of the world's most difficult problems will be found. Although we do not mean to denigrate the immediate problems facing us in the year 2003 and we sincerely would love to see peace in the Middle East, economic development in Central America, and a resolution to the war in the Congo, we insist that there are even deeper and more difficult problems associated, either directly or indirectly, with ecology. Indeed, there will be no final peace in the Middle East until the problem of access to water is resolved, an issue of sustainable resource use. Central America will not see economic development until the destruction of its natural environment ceases, a problem of ecology as much as of politics. And resolution of the conflict in the Congo will not completely solve the problem of epidemic diseases currently ravaging subsaharan Africa. These are all problems of ecology, and they will not be solved without attention to fundamental principles, the most elementary of which are the subject of this book.

Examples of practical applications have been peppered throughout the text, sometimes contrived, sometimes real. For the problems most obviously associated with the principles of population ecology we could force a classification based on what we see as emerging communities of scholars in contemporary society. First, for the effective utilization of natural resources, from water in the Middle East to land in Africa to petroleum in the world, the set of principles of resource consumption is an essential foundation. Chapters 1 and 3 specifically provide the foundations upon which most literature of natural resource harvesting is based (e.g., Getz and Haight 1989).

Second, the current crisis of biodiversity loss is ultimately related to the material presented in chapters 5, 6, and 8. The determination of what allows large numbers of species to coexist (Palmer 1994) and thus what needs to be modified to maintain all of those species, requires a knowledge of the material in those chapters. It is obvious that most of this problem will not be resolved without substantial political change, but it is also the case that even with utopian politics, human intervention will still threaten the world's biodiversity without a better understanding of what that biodiversity means in the first place.

Third, and somewhat more cryptically, there is a growing realization that the world is an uncertain place. The current revolution in nonlinear dynamics and complex systems has heightened our appreciation of the principle of "the inevitability of surprise" in any ecological system (Levins, personal communication). Indeed, much of the current appreciation of the uncertainty of complex dynamic systems was propelled by the work of an ecologist (May 1974). The material presented in chapter 4 is an attempt to introduce the student to the principles underlying this new field. Applications here are hard to predict, but practitioners are not averse to claiming grand breakthroughs in understanding. In application to the natural world, it promises to change at least some of the way we view environmental problems (e.g., Schaffer 1985, Hastings et al. 1993).

Fourth, there is an emerging field called ecosystems management. Probably most of the world's rural residents would be surprised to hear that we academics have just discovered what they and their ancestors have been doing for so many years, but so goes the academy. Thus, we are now concerned with the consequences of human activities on "ecosystem services." Here too, the principles introduced in this text are essential. For example, the management of agroecosystems has, for many years, required techniques for pest control, many of which relied on some sort of predator or parasite. The theory presented in chapter 6 is the underlying basis for the sometimes very sophisticated analytical approaches applied to this problem (Hawkins and Cornell 1999).

Fifth, who can ignore the fact that new diseases are emerging yearly and old diseases are emerging as new problems? The disaster of AIDS in Africa is fundamentally a problem of population ecology, the emergence of the Ebola virus likewise. Currently, there appears to be a spread of malaria into the East African highlands, likely due to environmental factors. To deal with these problems, in addition to the obvious public policy necessities, we need to more fully understand the epidemiology of each of the diseases. For this understanding, the material presented in chapter 7 is critical.

These five practical problems are intimately related to one another, and it would be foolish to suggest that the ecological principles involved in any way trump the political and economic barriers to their solution. However, each of them (and the collection of them together) contains problems that will not be solved, no matter how well the politics and economics are dealt with, without some attention to the underlying ecological principles. Introducing those principles is the purpose of this book. Students who master this material will be poised to delve into more sophisticated theoretical and empirical ecology and to understand some of the underlying basics needed to solve practical problems of the environment. We hope our presentation makes that mastery agreeable.

G l o s s a r y

---

attractor — An attractive region of the space defined by one or more variables. All trajectories not contained in that region will eventually wind up in the region. An attractor may be a point or a cycle that is an equilibrium and generates transients that return to the equilibrium state after perturbation. It may also be an attractive region that has no individual equilibrium points or cycles (a chaotic or strange attractor).

basin of attraction — The collection of points that converge on a particular attractor.

basin boundary collision — A type of bifurcation in which the edge of a basin of attraction intersects with the edge of a strange attractor.

bifurcation diagram — A graph of the attractors of a system as a function of some parameter (the "bifurcation" parameter).

bifurcation point — A point of structural instability in which a single equilibrium condition is split into two.

carrying capacity — The maximum attainable size of a population; usually symbolized as K.

chaos — Behavior of a system that is inherently unpredictable in the sense that two points of initiation that are extremely close together will generate trajectories that deviate from one another dramatically.

competitive exclusion principle — The idea, derived from the simple Lotka-Volterra equations, that if interspecific competition between two species is sufficiently large, the two both species cannot coexist in the same environment.

competitive production principle — The competitive exclusion principle applied to an agricultural situation, in which two crop species will overyield if the competition between them is sufficiently small.

density dependence — The condition wherein the rate at which a population increases or decreases is a function of its density (to be contrasted to density independence); often used interchangeably with the term *intraspecific competition* although not formally synonymous.

density independence — The condition where a population increases or decreases without relation to the density of the population (to be contrasted to density dependence).

elasticity — The degree to which the population growth rate changes as a function of changes in the elements of the projection matrix, expressed as a proportion.

equilibrium point — The value of a variable that does not change under the rules of a dynamic model. An equilibrium point may be stable (in which case it is commonly referred to as an attractor) or unstable (in which case it is commonly referred to as a repellor).

Euler's constant — Approximately 2.7183, the base of natural logarithms; normally symbolized with lower case e.

exponential function — Euler's constant raised to some value x, frequently symbolized as $\exp(x)$.

exponential growth — Pattern of increase (or decrease) in a population that follows the exponential equation, either the integrated form (equation 10 of chapter 1) or the differential equation form (equation 8 of chapter 1).

facultative mutualism — The case of mutualism in which one species can survive without its mutualist but performs better with it.

functional response — In consumer–resource (predator–prey) equations the function that stipulates how the per capita consumption rate (or predation rate) changes with changes in the resource density.

Gini coefficient — A measure of the inequality of a distribution of factors such as size or biomass.

intraspecific competition — The competitive interaction among individuals in the same population.

intrinsic rate of natural increase — The growth of a population under the theoretical state of extremely low population density; usually symbolized as r.

isocline — For a dynamic model, the set of all points for which one of the variables does not change (in the context of a differential equation, the set of points that corresponds to the derivative set equal to zero).

K *selection* — The selection regime that normally results in low reproduction potential and high survivorship, among other chatarcter-
istcs, as opposed to *r* selection, which normally results in higher reproductive potential and lower survivorship.

limit cycle — Oscillatory system that can be either stable (an oscillatory
attractor) or unstable (an oscillatory repellor).

logistic population growth — Population growth that appears qualitatively exponential at low population density but approaches an asymptote as the population gets larger. Population growth that follows the logistic equation (equation 17 in chapter 1).

metapopulation — A population distributed in patches in which each of
the patches is incapable of indefinitely maintaining a viable population alone (i.e., extinction probability > 0) but where the population
is maintained over the whole collection of patches because migrants
from occupied patches continually reoccupy patches in which subpopulations have gone extinct.

obligate mutualism — The case of mutualism in which one species is unable to survive without its mutualist.

one-dimensional map — A function that projects a single variable
through discrete time: e.g., $N_{t+1} = F(N_t)$.

paradox of enrichment — The tendency of a predator–prey system to become unstable with an increase in the environmental quality for the
prey.

population — A group of individual items. In the context of population
ecology, a population is a group of individual living organisms.

population projection matrix — The matrix of age-specific or stage-
specific transition probabilities and natality values that is used to
calculate the age- or stage-specific population densities into the
future.

propagule rain — In the context of a metapopulation, when all subpopulations release propagules generally into the environment such that
they "rain" on all subpopulations.

repellor — A point or cycle that is theoretically an equilibrium but generates transients that deviate from the equilibrium position when
perturbed.

reproductive value — The elements of the left eigenvector of a population projection matrix.

r *selection* — The selection regime that normally results in high reproduction potential and low survivorship, among other characteristics,

as opposed to *K* selection, which normally results in lower reproductive potential and higher survivorship.

rescue effect — In the context of a spatially subdivided population, when a subpopulation tends toward extinction but receives migrants from another subpopulation before going extinct.

sensitivity — The degree to which the population growth rate changes as a function of changes in the elements of the projection matrix.

separatrix — The boundary between two basins of attraction.

stable stage (age) distribution — The proportional distribution of individuals in age or stage classes after a population has been growing (with constant transition parameters) for very many generations. These proportions remain constant in perpetuity.

strange attractor — A chaotic attractor. Region of space that attracts all trajectories but contains no attractive points or cycles.

structural stability — A higher-level stability concept in which the qualitative nature of a system is unchanged when the parameters of the system are varied.

structured models — Models that do not assume all individuals in the population are identical.

thinning laws — The relationship between the population density and population biomass in a population undergoing mortality (thinning). Most often represented as a graph of the log of the biomass per individual versus the log of the density, which frequently results in a linear regression with a slope of −3/2, thus giving rise to the term three-halves thinning law.

vector field — The set of vectors that determine the behavior of a dynamic system.

yield–density relation — The yield (total biomass or the biomass of a part, e.g., seeds) of a population (especially a plant population) as a function of its density.

R e f e r e n c e s

Alligood, K. T., T. D. Sauer, and J. Yorke. 1996. *Chaos: An introduction to dynamical systems*. New York: Springer.

Altieri, M. A. 1987. *Agroecology: The scientific basis of alternative agriculture*. Boulder, CO: Westview Press.

Anderson, R. 1981. Population ecology of infectious disease agents. In *Theoretical ecology: Principles and applications*, edited by R. May. Sunderland, MA: Sinauer.

Anderson, R., and R. May. 1979. Population biology of infectious diseases: Part 1. *Nature* 280:361–367.

———. May. 1991. *Infectious diseases of humans: Dynamics and control*. London: Oxford Univ. Press.

Andrewartha, H. G., and L. C. Birch. 1954. *The distribution and abundance of animals*. Chicago: Univ. of Chicago Press.

Arditi, R., and A. A. Berryman. 1991. The biological control paradox. *Trends in Ecol. and Evol.* 6:32.

Armstrong, R. A., and R. McGehee. 1980. Competitive exclusion. *Am. Nat.* 115:151–170.

Ashman, T. L. 1994. A dynamic perspective on the physiological cost of reproduction in plants. *Am. Nat.* 144:300–316.

Bailey, N. T. J. 1975. *The mathematical theory of infectious diseases*. 2nd ed. London: Griffin.

Baltensweiler, W. 1971. The relevance of changes in the composition of larch bud moth populations for the dynamics of its numbers. In *Dynamics of populations*, edited by P. J. den Boer and D. R. Gradwell. Wageningen: Centre for Agricultural Publishing.

Batista, W. B., W. J. Platt, and R. E. Macchiavelli. 1998. Demography of a shade-tolerant tree *Fagus grandifolia* in a hurricane-disturbed forest. *Ecology* 79:38–53.

Blasius, B., A. Huppert, and L. Stone. 1999. Complex dynamics and phase synchronization in spatially extended ecological systems. *Nature* 399:354–359.

Bleasdale, J. K. A., and J. A. Nelder. 1960. Plant population and crop yield. *Nature*, Lond. 188:342.

Boucher, D. H., and Mallona, M. A. 1997. Recovery of the rain forest tree *Vochysia ferruginea* over 5 years following Hurricane Joan in Nicaragua: A preliminary population projection matrix. *For. Ecol. Manage.* 91:195–204.

Boyce, M. S. 1984. Restitution of r- and K-selection as a model of density-dependent natural selection. *Annu. Rev. Ecol. Syst.* 15:427–447.

Branch, G. M. 1975. Intraspecific competition in *Patella cochlear Born. J. Anim. Ecol.* 44:263–281.

Caswell, H. 1978. A general formula for the sensitivity of population growth rate to changes in life history parameters. *Theor. Popul. Biol.* 14:215–230.

———. 2001. *Matrix population models. Construction, analysis and interpretation.* Sunderland, MA: Sinauer.

Caswell, H., R. Naiman, and R. Morin. 1984. Evaluation of the consequences of reproduction in complex salmonid life cycles. *Aquaculture* 43:123–143.

Chapman, R. 1928. The quantitative analysis of environmental factors. *Ecology* 9:111–122.

Cody, M. L. 1966. A general theory of clutch size. *Evolution* 20:174–184.

Cole, L. C. 1954. The population consequences of life history phenomena. *Quart. Rev. Biol.* 29:103–137.

Cornell, H. V. 1985. Species assemblages of cynipid gall wasps are not saturated. *Am. Nat.* 126:565–569.

Costantino, R. F., R. A. Desharnais, J. M. Cushing, and B. Dennis. 1997. Chaotic dynamics in an insect population. *Science* 275:389–391.

Deevey, E. S., Jr. 1947. Life tables for natural populations of animals. *Quart. Rev. Biol.* 22:283–314.

de Kroon, H., A. Plaisier, J. van Groenendael, and H. Caswell. 1986. Elasticity: The relative contribution of demographic parameters to population growth rate. *Ecology* 67:1427–1431.

Dempster, J. P. 1983. The natural control of populations of butterflies and moths. *Biol. Rev.* 58:461–481.

Diggle, P. J. 1983. *Statistical analysis of spatial point patterns.* New York: Academic Press.

Durrett, R., and S. Levin 1994. Stochastic spatial models: A user's guide to ecological applications. *Philos. Trans. R. Soc. London ser. B* 343:329–350.

———. 1998. Spatial aspects of interspecific competition. *Theor. Pop. Biol.* 53: 30–43.

Ellner, S., and P. Turchin. 1995. Chaos in a noisy world: New methods and evidence from time-series analysis. *Am. Nat.* 145:343–375.

Eriksson O., and K. Kiviniemi. 1999. Site occupancy, recruitment and extinction thresholds in grassland plants: An experimental study. *Biol. Conserv.* 87: 319–325.

Fan, Y. Q., and F. L. Petitt. 1994. Parameter estimation of the functional response. *Environ. Entomol.* 23:785–794.

Futuyma, D. J. 1979. *Evolutionary biology*. Sunderland, MA: Sinauer.

Gause, G. F. 1934. *The struggle for existence*. Baltimore: Williams and Wilkins.

Getz, W. M., and R. G. Haight. 1989. *Population harvesting: Demographic models of fish, forest and animal resources*. Princeton Monographs in Population Biology 27. Princeton, NJ: Princeton Univ. Press.

Gotelli, N. J. 1991. Metapopulation models: The rescue effect, the propagule rain and the core-satellite hypothesis. *Am. Nat.* 138:768–776.

Gross, K. L. 1984. Effects of seed size and growth form on seedling establishment of six monocarpic perennial plants. *J. Ecol.* 72:369–387.

Gukenheimer, J., G. Oster, and A. Ipaktchi. 1977. The dynamics of density dependent population models. *J. Math. Biol.* 4:101–147.

Hairston, N. G., D. W. Tinkle, and H. M. Wilbur. 1970. Natural selection and the parameters of population growth. *J. Wildl. Mange.* 34:681–690.

Hanski, I. 1982. Dynamics of regional distributions: The core and satellite species hypothesis. *Oikos* 38:210–221.

———. 1999. *Metapopulation biology: Ecology, genetics, and evolution*. San Diego: Academic Press.

Hanski, I., J. Kouki, and A. Halkka. 1983. Three explanations of the positive relationship between distribution and abundance of species. Pages 108–116 in *Community diversity: Historical and geographical perspectives*, edited by R. E. Ricklets, and D. Schluter. Chicago: Univ. of Chicago Press.

Hanski, I., and D. Simberloff. 1997. The metapopulation approach, its history, conceptual domain and application to conservation. In *Metapopulation biology: ecology, genetics, and evolution*, edited by I. Hanski, and M. Gilpin. San Diego: Academic Press.

Harrison, S. 1991. Local extinction in a metapopulation context. In *Metapopulation dynamics: empirical and theoretical investigations*, edited by I. Hanski and M. Gilpin. London: Academic Press.

Harrison, S., and A. D. Taylor. 1997. Empirical evidence for metapopulation dynamics. In *Metapopulation biology: ecology, genetics, and evolution*, edited by I. Hanski and M. Gilpin. San Diego: Academic Press.

Hassell, M. P. 1978. *The dynamics of arthropod predator-prey systems*. Princeton Monographs in Population Biology. Princeton, NJ: Princeton Univ. Press.

Hastings, A., C. L. Hom, S. Ellner, P. Turchin, and H. C. J. Godfray. 1993. Chaos in ecology: Is mother nature a strange attractor? *Ann. Rev. Ecol. Syst.* 24:1–33.

Hastings, A., and T. Powell. 1991. Chaos in a three species food chain. *Ecology* 72:896–903.

Hawkins, B. A., and H. V. Cornell. 1999. *Theoretical approaches to biological control*. Cambridge: Cambridge Univ. Press.

Hiroi, T., and M. Monsi. 1966. Dry-matter economy of *Helianthus annus* communities grown at varying densities and light intensities. *J. Fac. Sci. Univ. Tokyo* 9:241–285.

Holling, C. S. 1959. Some characteristics of simple types of predation and parasitism. *Can. Entomol.* 91:385–398.

Huffaker, C. B. 1958. Experimental studies on predation: Dispersion factors and predator-prey oscillations. *Hilgardia* 27:343–383.

Jackson, E. A. 1991. *Perspectives of nonlinear dynamics*. Vols. 1 and 2. Cambridge: Cambridge Univ. Press.

Jackson, L. L., and C. L. Dewald. 1994. Predicting evolutionary consequences of greater reproductive effort in *Tripsacum dactyloides*, a perennial grass. *Ecology* 75:627–641.

Jackson, W. 1980. *New roots for agriculture*. San Francisco: Friends of the Earth.

Kays, S., and J. L. Harper. 1974. The regulation of plant and tiller density in a grass sward. *J. Ecol.* 62:97–10

Lack, D. 1947. The significance of clutch-size. Pts. 1 and 2. *Ibis* 89:302–352.

Lande, R. 1982. A quantitative genetic theory of life history evolution. *Ecology* 63:607–615.

Lefkovitch, L. P. 1965. The study of population growth in organisms grouped by stages. *Biometrics* 21:1–18.

Leslie, P. H. 1945. On the use of matrices in certain population mathematics. *Biometrika* 33:183–212.

Levins, R. 1968. *Evolution in changing environments*. Princeton, NJ: Princeton Univ. Press.

———. 1969. Some demographic and genetic consequences of environmental heterogeneity for biological control. *Bull. Entomol. Soc. Am.* 15:237–240.

———. 1974. Qualitative analysis of partially specified systems. *Ann. NY Acad. Sci.* 231:123–138.

———. 1979. Coexistence in a variable environment. *Am. Nat.* 114:765–783.

Levins, R., and J. H. Vandermeer. 1990. The agroecosystem embedded in a complex ecological community. Pages 341–362 in *Agroecology*, edited by C. R. Carroll, J. H. Vandermeer, and P. Rosset. New York: McGraw-Hill.

Lewis, E. G. 1942. On the generation and growth of a population. *Sankhyaa* 6:93–96.

Lewontin, R. C. 1969. The meaning of stability. *Brookhaven Symp. Biol.* 22:13–24.

Linkola, K. 1916. Studien Über den Einflus der Kultur auf die Flora in den Gegenden nordlich von Ladogasee. I. Allgemeiner Teil. *Acta Soc. Fauna Flora Fenn.* 45:1–432.

Lotka, A. J. 1926. *Elements of physical biology*. Baltimore: Williams and Wilkins.

MacArthur, R. 1970. Species packing and competitive equilibria for many species. *Theor. Popul. Biol.* 1:1–11.

MacArthur, R. H., and E. O. Wilson. 1967. *The theory of island biogeography.* Princeton, NJ: Princeton Univ. Press.

Macdonald, G. 1957. *The epidemiology and control of malaria.* London: Oxford Univ. Press.

Malthus, T. 1960. *A Summary View of the Principle of Population* (originally published in 1830). New York: New American Library.

May, R. M. 1973. *Stability and complexity in model ecosystems.* Monographs in Population Biology. Princeton, NJ: Princeton Univ. Press.

———. 1974. Biological populations with nonoverlapping generations: Stable points, stable cycles and chaos. *Science* 186:645–647.

———. 1977. Thresholds and breakpoints in ecosystems with a multiplicity of stable states. *Nature* 269:471–477.

———. 1981. *Theoretical ecology: Principles and applications.* 2nd ed. Sunderland, MA: Sinauer.

McCann, K., and P. Yodzis. 1994. Nonlinear dynamics and population disappearances. *Am. Nat.* 144:873–879.

Milne, A. 1961. Definition of competition among animals. In Mechanisms in Biological Competition, edited by F. L. Milthorpe. *Symp. Soc. Exp. Biol.* 15:40–61.

Morales, H. 1998. Indigenous methods of pest control in the Guatemalan highlands. Ph.D. diss. Univ. of Michigan, Ann Arbor.

Murdoch, W. W., S. L. Swarbrick, R. F. Luck, S. Walde, and D. S. Yu. 1996. Refuge dynamics and meta population dynamics: An experimental test. *Am Nat.* 147:424–444.

Murie, O. 1944. *The wolves of Mt. McKinley.* Fauna Series, No. 5. Washington, DC: U.S. Department of the Interior, National Park Services.

Nicholson, A. J. 1933. The balance of animal populations. *J. Anim. Ecol.* 2:131–178.

———. 1957. The self-adjustment of populations to change. *Cold Spring Harbor Symp. Quant. Biol.* 22:153–173.

Nicholson, A. J., and V. A. Bailey. 1935. The balance of animal populations. *Proc. Zool. Soc. Lond.* 1:551–598.

Noy-Meir, E. 1975. Stability of grazing systems: An application of predator-prey graphs. *J. Ecol.* 63:459–481.

Paige, K. N., and T. G. Whitham. 1987. Overcompensation in response to mammalian herbivory: The advantage of being eaten. *Am. Nat.* 129:407–416.

Palmer, M. W. 1994. Variation in species richness: Towards a unification of hypotheses. *Folia Geobot. Phytotaxon. Praha.* 29:511–530.

Pascual, M., and S. A. Levin. 1999. From individuals to population densities:

Searching for the intermediate scale of nontrivial determinism. *Ecology* 80:2225–2236.

Pascual, M., P. Mazzega, and S. Levin. 2001. Oscillatory dynamics and spatial scale: The role of noise and unresolved pattern. *Ecology* 82:2357–2369.

Pearl, R., and L. J. Reed. 1920. On the rate of growth of the population of the United States since 1790 and its mathematical representation. *Proc. Nat. Acad. Sci. USA* 6:275–288.

Pianka, E. 1970. On r and K selection. *Am. Nat.* 104:592–597.

Pollard, E., K. H. Lakhani, and P. Rothery. 1987. The detection of density dependence from series of annual censuses. *Ecology* 68:2046–2055.

Reznick, D. 1985. Costs of reproduction: An evaluation of the empirical evidence. *Oikos* 44:257–267.

———. 1992. Measuring the costs of reproduction. *Trends in Ecol. and Evol.* 7:42–45.

Risch, S. J., and D. H. Boucher. 1976. What ecologists look for. *Bull. Ecol. Soc. Am.* 57:8–9.

Roff, D. A. 1992. *The evolution of life histories: Theory and analysis*. New York: Chapman and Hall.

Rose, M. R., and B. Charlesworth. 1981. Genetics of life history in *Drosophila melanogaster*. II. Exploratory selection experiments. *Genetics* 97:187–196.

Rose, M. R., P. M. Service, and E. W. Hutchinson. 1987. Three approaches to trade-offs in life-history evolution. Pages 93–105 in *Genetic constraints on adaptive evolution*, edited by V. Loeschcke. Berlin: Springer-Verlag.

Rosenzweig, M. L. 1971. Paradox of enrichment: Destabilization of exploitation ecosystems in ecological time. *Science* 171:385–387.

Rosenzweig, M. L., and R. H. MacArthur. 1963. Graphic representation and stability conditions of predator-prey interactions. *Am. Nat.* 97:209–223.

Ross, R. 1911. *The prevention of malaria*. 2nd ed. London: Murray.

Schaffer, W. M. 1974. Optimal reproductive effort in fluctuating environments. *Am. Nat.* 108:783–790.

———. 1985. Order and chaos in ecological systems. *Ecology* 66:93–106.

Schaffer, W. M., and M. D. Gadgil. 1975. Selection for optimal life histories in plants. In *Ecology and evolution of communities*, edited by M. L. Cody and J. M. Diamond. Cambridge: Harvard Univ. Press.

Schwinning, S., and J. Weiner. 1998. Mechanisms determining the degree of size asymmetry in competition among plants. *Oecologia* 113:447–445.

Shinozaki, K., and T. Kira. 1956. Intraspecific competition among higher plants. VII. Logistic theory of the C-D effect. *J. Inst. Polytech. Osaka City Univ.* 7:35–72.

Silvertown, J., and M. Dodd. 1996. Comparing plants and connecting traits. In *Plant life histories*, edited by J. Silvertown, M. Franco, and J. L. Harper. Cambridge: Cambridge Univ. Press.

Smith, H. S. 1935. The role of biotic factors in the determination of population densities. *J. Econ. Entomol.* 28:873–898.

Solé, R., and B. Goodwin. 2000. *Signs of life: How complexity pervades biology.* New York: Basic Books.

Soule, J., D. Carré, and W. Jackson. 1990. Ecological impact of modern agriculture. Pages 165–188 in *Agroecology,* edited by C. R. Carroll, J. H. Vandermeer, and P. Rosset. New York: McGraw-Hill.

Stearns, S. 1992. *The evolution of life histories.* Oxford: Oxford Univ. Press.

Stock, T. M., and J. C. Holmes. 1988. Functional relationships and microhabitat distributions of enteric helminthes of grebes (Podicipedidae): The evidence for interactive communities. *J. Parasitol.* 74:844–856.

Stiven, A. E. 1967. The influence of host population space in experimental epizootics caused by *Hydramoeba hydroxena. J. Invertebr. Pathol.* 9:536–545.

Strong, D. 1986. Density-vague population change. *Trends in Ecol. and Evol.* 1:39–42.

Thompson, W. R. 1928. A contribution to the study of biological control and parasite introduction in continental areas. *Parasitology* 20:90–112.

Tilman, D. 1982. *Resource competition and community structure.* Princeton, NJ: Princeton Univ. Press.

———. 1997. Community invasibility, recruitment limitation and grassland biodiversity. *Ecology* 78:81–92.

Turner, M. D., and D. Rabinowitz. 1983. Factors affecting frequency distributions of plant mass: The absence of dominance and suppression in *Festuca paradoxa. Ecology* 64:469–475.

Utida, S. 1957. Population fluctuation: An experimental and theoretical approach. *Cold Spring Harbor Symp. Quant. Biol.* 22:139–151.

Vandermeer, J. H. 1969. The competitive structure of communities: An experimental approach using protozoa. *Ecology* 50:362–371.

———. 1975. On the construction of the population projection matrix for a population grouped in unequal stages. *Biometrics* 31:239–242.

———. 1984. Plant competition and the yield-density equation. *J. Theor. Biol.* 109:393–399.

———. 1989. *The ecology of intercropping.* Cambridge: Cambridge Univ. Press.

———. 1997. Period "bubbling" in simple ecological models: Pattern and chaos formation in a quartic model. *Ecol. Model.* 95:311–317.

Vandermeer, J. H., and D. Boucher. 1978. Varieties of mutualistic interaction in population models. *J. Theor. Biol.* 74:549–558.

Vandermeer, J. H., and R. Carvajal. 2001. Metapopulation dynamics and the quality of the matrix. *Am. Nat.* 158:211–220.

Vandermeer, J. H., B. Schultz, P. Rosset, H. McGuinnes, and I. Perfecto. 1984. An ecologically-based approach to the design of intercrop agroecosystems:

An intercropping system of soybeans and tomatoes in southern Michigan. *Ecol. Model.* 25:121–150.

Vandermeer J. H., and P. Yodzis. 1999. Basin boundary collision as a model of discontinuous change in ecosystems. *Ecology* 80:1817–1827.

Verhulst, P. F. 1838. Notice sur la loi que la population suit dans son accrioissement. *Corresp. Math. Phys.* 10:113–121.

Volterra, V. 1926. Variazioni e fluttuazioni del numero d'individui in specie animali conviventi. *Mem. Acad. Lincei* 2:31–113.

Warren, K. S., and V. A. Newill. 1967. *Schistosomiasis—A bibliography of the world's literature from 1852 to 1962.* Vols. 1 and 2. Cleveland, OH: Western Reserve Univ. Press.

Weiner, J. 1986. How competition for light and nutrients affects size variability in *Ipomoea tricolor* populations. *Ecology* 67:1425–1427.

———. 1990. Asymmetric competition in plant populations. *Tree* 5:360–364.

Weiner, J., and S. C. Thomas. 1986. Size variability and competition in plant monocultures. *Oikos* 47:211–222.

Westoby, M. 1984. The self-thinning rule. *Adv. Ecol. Res.* 14:167–225.

Wiens, J. A., N. C. Stenseth, and B. VanHorne. 1993. Ecological mechanisms and landscape ecology. *Oikos* 66:369–380.

Wilbur, H. M., D. W. Tinkle, and J. P. Collins. 1974. Environmental certainty, trophic level, and resource availability in life history evolution. *Am. Nat.* 108:805–817.

Willey, R. W., and S. B. Heath. 1969. The quantitative relationships between plant population and crop yield. *Adv. Agron.* 21:281–321.

Yoda, K., T. Kira, H. Ogawa, and H. Hozumi. 1963. Self-thinning in overcrowded pure stands under cultivated and natural conditions. *J. Inst. Polytech. Osaka City Univ. Ser. D.* 14:107–129.

Yodzis, P. 1978. *Competition for space and the structure of ecological communities.* Berlin: Springer-Verlag.

Index